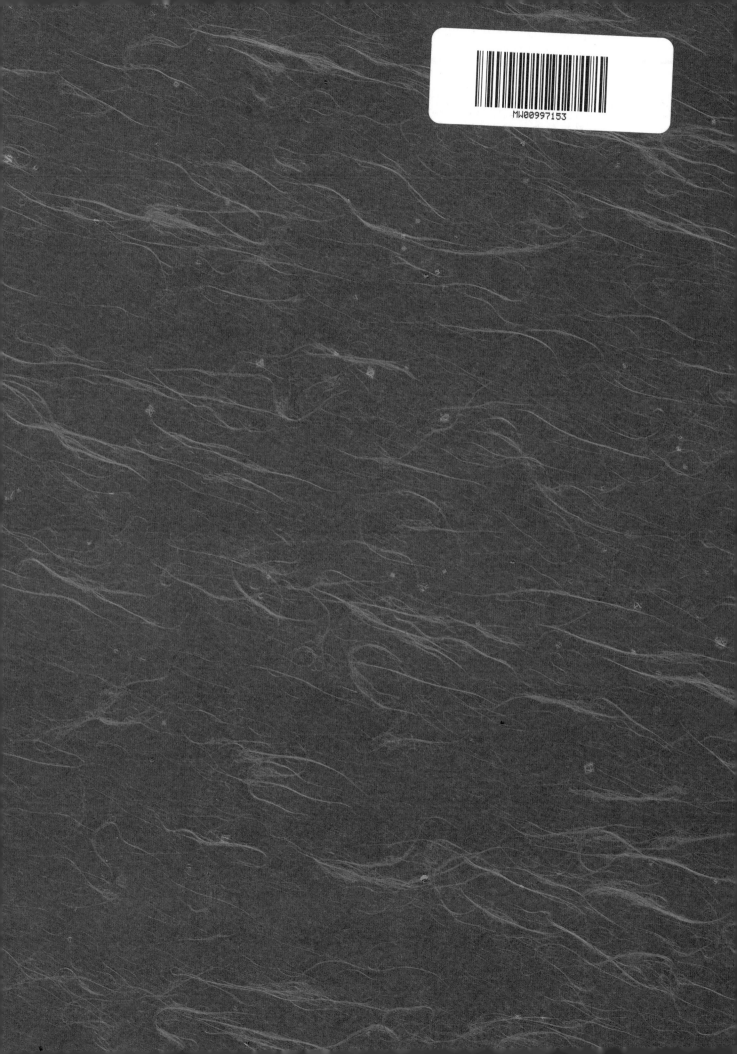

ZIPPO
The Great American Lighter

David Poore

Revised and Expanded 2nd Edition

Including the Poore Guide to Zippo Prices

4880 Lower Valley Road, Atglen, PA 19310 USA

Dedication

To my family:

My parents and brother, who provide the flint
My sons, John, Mark, Matthew, Luke, and Caleb
Who add the spark
And to my wife Mary, who tends the flame

Revised and Expanded 2nd Edition
Copyright © 2005 by David Poore
Originally copyrighted © in 1997
Library of Congress Control Number: 2005924561

All rights reserved. No part of this work may be reproduced or used in any form or by any means—graphic, electronic, or mechanical, including photocopying or information storage and retrieval systems—without written permission from the publisher.
The scanning, uploading and distribution of this book or any part thereof via the Internet or via any other means without the permission of the publisher is illegal and punishable by law. Please purchase only authorized editions and do not participate in or encourage the electronic piracy of copyrighted materials.
"Schiffer," "Schiffer Publishing Ltd. & Design," and the "Design of pen and ink well" are registered trademarks of Schiffer Publishing Ltd.

Designed by "Sue"
Type set in Souvenir

ISBN: 0-7643-2337-7
Printed in China
1 2 3 4

Published by Schiffer Publishing Ltd.
4880 Lower Valley Road
Atglen, PA 19310
Phone: (610) 593-1777; Fax: (610) 593-2002
E-mail: Info@schifferbooks.com

For the largest selection of fine reference books on this and related subjects, please visit our web site at
www.schifferbooks.com
We are always looking for people to write books on new and related subjects. If you have an idea for a book please contact us at the above address.

This book may be purchased from the publisher.
Include $3.95 for shipping.
Please try your bookstore first.
You may write for a free catalog.

In Europe, Schiffer books are distributed by
Bushwood Books
6 Marksbury Ave.
Kew Gardens
Surrey TW9 4JF England
Phone: 44 (0) 20 8392-8585; Fax: 44 (0) 20 8392-9876
E-mail: info@bushwoodbooks.co.uk
Free postage in the U.K., Europe; air mail at cost.

Author's Note: The author both buys and sells common and rare Zippo lighters on a regular basis. His company is Zippo-Mania. The fax number is (308) 784-4684 and his home phone number is (308) 784-3184. Address: 1615 B Street, Cozad, NE 69130. Email: dpoore@cozadtel.net or dpoore@hotmail.com

CONTENTS

Acknowledgments 4
Foreword 5
Introduction 7
Early History 9
Zippo Chronology (Milestones) 14
    The Lighters 17
    The First Models 17
    Sports Series 37
    Town and Country Series 65
    The War and Beyond 76
    Special Interests 91
    The Barrett-Smythe Collection:
        1993-1994 113
    Table Lighters 129
    Zippo Accessories 139
    Military Lighters 144
    Desirable Advertisers 149
    Miscellaneous Zippo Products 171
Zippo Glossary 178
Dating Zippos 182
    Regular Size Cases 182
    Slim Size Cases 193
    Regular Size Inserts from 1955 to Date 194
Slim Inserts 197
The Poore Guide to Zippo Prices 202
Bibliography 215

ACKNOWLEDGMENTS

There are many people whose advice and assistance were invaluable in the production of this book. I would particularly like to acknowledge Hiroshi Kito, without whose help this book would be neither as complete nor as accurate. I would especially like to thank him for allowing me to use his charts. I would like to thank the people of Zippo Manufacturing Company, past and present. In my research, I interviewed a number of early Zippo employees. These men and women are priceless sources of oral history, pieces of Americana themselves. I am proud to have been able to hear and help preserve some of their knowledge. They include Bette Ross, George Duke, who is the grandson of George Blaisdell, Rudy Bickel, Joe Fearn, Mrs. Dale Hutton, Lillian Mezzeio, Bob Galey, who was former president and sales manager, Howard Fesenmyer, who is now executive secretary of the Blaisdell Foundation, Mike Pehonsky, and the late Wayne Edwards, whose early history of Zippo has been useful. I would like to thank some of the current Zippo employees, including Peggy Errera, Mike McLaughlin, who photographed many of the lighters in the text, Pat Grandy, and Julio Pedine. I also thank Chet Smith, whose help was invaluable, Jerry Burkhardt, Tim Ziaukas, Ted Ballard, Bill Kim, Ira Pilossof, Brian Sipes, Terry Cairo, Don Collier, Bucky Clinger, Gary Sneary, Ed Fingerman, Sam Wood, Keith Zoransky, Steve and Dot Terry, and Mel Muntzing for their help, encouragement and friendship. Finally, I thank my son, Mark Poore, for his graphic work throughout the book. Any errors, however, which may have crept into the text despite my best efforts, are entirely my own.

FOREWORD

Why a book on Zippo lighters? The answer may not be as simple as it seems.

Yes, there are hundreds of thousands of lighter collectors around the world, and many of them specialize in the windproof lighter made in Bradford, Pennsylvania. This year's annual Zippo/Case International Swap meet, held every July in Bradford, attracted 5,000 participants. So a book that offers the history of Zippo lighters, complete with a method to identify, date, and even price these bits of Americana, is useful.

But the pioneering work that David Poore has done here is, I believe, more important than that, and his efforts will outlive the transitory use of a company history or a price guide.

Why? Because Zippo lighters mean something.

The Zippo lighter represents a moment in the history of technology, and we need to look to cultural history—not the antique or collectible market—to help us understand this instrument and the social context in which it was developed and used. Poore's work opens the door to this inquiry.

Poore recounts the history of Zippo Manufacturing Company founded by George G. Blaisdell in 1932. It is, of course, one of the great American success stories, a tale of spunk and innovation, of fortune making and of the emergence of an iconic American product, a name as enduring as Coca-Cola, Levis, and Xerox.

But more importantly, Poore identifies and details the production history of the various models and styles of Zippo lighters. From the early 1932-33 tall models, which today can sell for over $30,000 on the collectors' market in mint condition, to some of the current models produced at a rate of 80,000 a day, Poore marks the highlights of the more than three hundred million Zippo lighters produced at the Bradford site and then shipped throughout the world.

These lighters—smooth fire starters that fit so snugly in the human palm and snap shut with such comfortingly firm and distinct finality—have become a ubiquitous part of American popular culture. In many cases, they have been a canvas on which companies, celebrities, events and even an individual's hopes and fears have been etched, baked, even scratched and gouged. Lighter historians Stuart Schneider and George Fischler write "like small clay tables of ancient Egypt, [Zippo lighters] will tell social history thousands of years into the future." They are right.

Robert Friedel, who teaches the history and technology at the University of Maryland, writes in his history of the zipper, an invention after which the Zippo lighter was named: "It is in part just because the zipper [and I would add the Zippo lighter] has become an invisible but inescapable part of daily life in the mid- and late-twentieth century that it is such an appropriate subject for an exercise in what the Swiss critic Sigfried Giedion called 'anonymous history'."

Zippo lighters are, in fact, the slates on which we write what is on collective mind. It remains for cultural historians to read these signs and symbols. On these lighters is written the history of advertising and public relations—arguably two of the archetypal endeavors of our time. The lighters tell the tales of the great wars, as well as the images and trends, sometimes historic, sometimes kitschy, that have riveted our attention or caught our fancy. The images and icons that have bubbled up and been fixed on these fire-starters exist as a kind of cultural Rorschach: Elvis and Mickey Mouse, smiley faces and presidents, pinups and rain forests. As current Zippo president Michael Schuler correctly says: "Every lighter is a story."

Historian Friedel would, I believe, agree. For "one of the goals of this kind of [anonymous] history," he writes, "is to understand better what the things that we make and use in our daily lives reveal about ourselves—value, ambitions, intentions, and prejudices that we may not even be conscious of."

Poore's work allows us to commence with the task of deciphering these articles in their social context, a context that goes back long before the advent of history and into the mythic past.

Zippo lighters, with their attached lid and works-every-time guarantee, bring the story of fire making and tending into the modern world in a particular, specialized way. For it is with the Zippo lighter, as its worldwide marketing hegemony as a result of World War II, fire making could be done on a large scale with that iconic gesture of our time: the flick of a finger. Zippo lighters incorporate that archetypal modern gesture—that push-of-a-button instantaneousness—with that most ancient need: to keep and tend the fire.

As a result, I expect Poore's book to spark interest in areas far beyond the antique stores, flea markets, and swap meets. His work will allow us to examine that moment in the social history of technology when humankind tamed the prize of Prometheus and started to carry it around in its pocket.

Tim Ziauka
Assistant Professor of Public Relations
University of Pittsburgh at Bradford
26 August 1996

INTRODUCTION

As a child, I was always intrigued with miniature lift-arm cigarette lighters that came in gumball machines. My parents indulged me by allowing me to buy them as souvenirs on family vacations.

With time my interests evolved and I began collecting turn-of-the-century lighters that were ignited by caps made of mercury fulminate. It was not until I was invited by my good friend, Don Collier, to attend a lighter convention that my interests changed solely to Zippo lighters. At the convention that he hosted, I saw such enthusiasm over Zippo lighters that I, too, found myself caught up in the zeal.

I was deeply impressed by the Japanese collectors who attended the show. Their interest in Zippo lighters seemed unbounded. Their enthusiasm fueled mine. Japanese buyers set out to buy every available Zippo lighter. This too inflamed my passion. Seeing the opportunity to make money I began to purchase Zippo lighters with the idea of entering the Japanese market. Throughout my dealings with the Japanese, many rare and unusual Zippo lighters enamored me. I couldn't bring myself to part with many of the rarest pieces whatever the price. This led to serious collecting and, eventually, to recognizing the need for a book like this. In only seven short years I have amassed the largest rare Zippo lighter collection in the United States. My collection is mainly comprised of prototypes and test models for many of the series that Zippo has offered since 1932.

As my interests grew and the interest for Zippo lighters in the world market grew keen the need for this book grew more apparent. As a result this book contains a sequential list of most series models with access numbers, history, illustrations, and both a price and rarity guide. My collection contains not only prototypes and test models, as you can see, but also a number of the elusive 1932/1933 models. I also own the largest Town and Country Sports model collection, which is considered by many to be the most beautiful and desirable series to ever be produced by Zippo.

Interest in Zippo lighters has burgeoned and the vast majority of new lighter collectors collect Zippo lighters. The market for Zippo lighters is stronger today than ever before. Zippo-mania is not only sweeping the United States, but the world as well. I hope that it engulfs you and you come to enjoy it as much as I have.

This book contains a sequential history of series, cases, inserts, fluid cans, flint packages, and sundries. It also includes many of the most highly prized Zippo lighters that people collect. Information was based on original Zippo salesman's catalogs, leaflets, advertising brochures, and the study of thousands of Zippo lighters. For various reasons this book does not contain every series or outstanding illustration that Zippo has ever produced; neither does every model with an access number, have an accompanying picture. It does however, contain "most" of the more significant series, as well as many of the most desirable Zippo lighters from this author's perspective. Prices are provided with each picture, and a complete price guide is located in the back of the book. Note: All lighters are valued in excellent condition, regardless of the condition in the photograph. Excellent condition will mean that a lighter has all of its original parts unless otherwise stated.

THE EARLY YEARS OF ZIPPO HISTORY (1932-1941)

The story of Zippo's beginning that seemed to be accurate, has slightly changed, due to new found information. The story initially went something like this. George Blaisdell was on the terrace of a country club in Bradford one summer evening in 1932. A gentleman was on the terrace with him trying to light his cigarette with an awkward Austrian lighter. Blaisdell wasn't real impressed with the lighter, although, after a short period of time, he got the patent rights for the Austrian lighter. He made some minor improvements, such as chrome plating it, but the lighter didn't sell well and needed more modifications. Blaisdell then decided to use his machinist skills to fabricate a new lighter which he gave the name, Zippo.

The incident is true but didn't take place in 1932. The summer incident actually took place the summer of 1931. Blaisdell then got the Austrian patent, imported units, made modifications, and then tried to sell the new product.

During the first month, he produced 82 units with sales totaling $69.15 (retailing at $1.95 / unit). It was believed by both Zippo and collectors that Blaisdell was talking about his initally believed 1932, tall square model. Actually the 82 units he referred to were his modified Austrian models, which accounts for the perceived descrepancies. Blaisdell had even named his new modified Austrian lighter, Zippo which enhanced the confusion by both collectors and Zippo employees some 65 years later.

According to Zippo documents and personal letters between Blaisdell and the Austrian Co., Blaisdell didn't begin actual full scale production of the taller 1933 models until February, 1933. By the end of 1933, Blaisdell reduced the size of the original lighter by 1/4 inch. Therefore, Blaisdell only produced the 1/4 inch taller models with figure eight cams during 1933. He made two distinct styles of tall models. He made one variation without diagonal lines as well as one with diagonal lines. I have given the 1932/1933 tall model without lines, access #1 whereas I have given the ones with diagonal lines, access #2.

According to tangible records Blaisdell founded his company, "Zippo" in 1932. Some cite the fact that Blaisdell did not apply for the patent until 1934, as proof of a later beginning for Zippo. As stated before in my first book, since "Pat. Pending" appears on the earliest Zippo lighters it is argued that they must date to 1934. Ted Ballard and others pointed out, that it was a matter of professional policy to stamp "patent pending" when fabricating a new product whether the patent was actually applied for or not, in order to "try to prevent" others from pirating the idea. This was especially true on metal items.

Blaisdell both fabricated and marketed the Zippo lighter long before he applied for the patent. One obvious reason was that he wanted to see if his "Zippo" would sell. Why continue if there is no market? During this interim period he needed to sell lighters both to to make a living and keep the company solvent. These were tough times not only for Blaisdell, but also the rest of the country.

According to Zippo sources, the "tall" model 1933's were only made for the first quarter of 1933 and then Zippo switched production to the ¼ inch shorter units for the latter part of 1933 and on into 1936 (with the hinge plates still on the outside).

Again, given the current available data, 1932/33 tall models were only produced for a few months. The production process was slow due to using hot plates for soldering, so it is more than likely that not many 1932 or 1933 tall models were ever produced. This is all speculative, but about as good as can be expected without tangible records. Current data suggests that there were as few as 600 and as many as 1,500 tall model units made in 1933.

An extremely rare Coca-Cola metallique. It is one of only two known to exist.

Zippos were introduced at the height of the Great Depression. Their retail price of $1.95 was a considerable sum during the 1930s. The "lipstick" style lighters that were produced during that same time frame sold for as little as fifteen cents. The consumer who paid $1.95 for a lighter would not have stored it away in some drawer. Few people could afford doing that. Rather they would have used it until it literally wore out.

One of the reasons that so few are known to exist is that the soldered spots on the lighter tended to break with prolonged use, especially on the case. The vast majority of the models that still exist have hairline fractures along the soldered sites. Most 1932/1933 tall units have been repaired, in some way, by Zippo. Zippo's lifetime warranty allows for worn out lighters to be sent to Zippo to be repaired or replaced with a new lighter at Zippo's discretion. It is not known how many are extant. Other than those in the Zippo museum, twenty-nine 1932/1933 tall models are known to exist in the United States. I, personally, know of ten 1932/1933's that exist in Japan. Only two 1932/1933's are known to exist in Europe, according to lighter club members in the various international clubs. Thus, only forty-one 1932/1933 tall models are known to exist in any condition throughout the world. Fewer than one-fifth of those have totally original parts. Only seven of the forty-one 1932/33 tall models have the original "hook cam."

How can we guarantee every Zippo to work forever?

Because we build them to stay out of the repair shop. If they ever need adjustment, your customers just send them to us to be put in perfect working order, without a cent of cost to them . . . or trouble to you!

How many other things do you sell that give your customers this kind of guarantee?

One reason why we dare to do it is pictured below—an exclusive Zippo construction detail. Show your customers! What better gift for any man or woman smoker than a beautiful Zippo, with its sensational guarantee? Zippos give dealers a break too, through fast turnover, a good profit and satisfied customers.

When you are asked to handle an unknown or fly-by-night imitation of Zippo, remember that a solid business is built on satisfied customers . . . that performance is remembered long after price is forgotten.

No "flopover flints" because the flint fits into this hard steel collar at the top of flint tube

ZIPPO'S PATENTED CONSTRUCTION PREVENTS THIS

Lighters give trouble—bring customers running back to the dealer—when the flint wears against the metal tube. The loose flint then catches under the spark wheel and jams. But that can't happen with Zippo. The patented hard steel collar holds the flint straight and firm against the wheel for a lifetime of sure, easy lights.

A "flopover flint." Note how the hard flint has worn the soft metal tube lopsided. This can mean a jammed, useless lighter.

ZIPPO GUARANTEED TO WORK FOREVER
ZIPPO MANUFACTURING COMPANY, BRADFORD, PA.

If any Zippo® Product ever fails to work... we'll fix it free

This is your assurance that no Zippo customer has ever been, nor will he ever need to be, unhappy with the products and repair policy of Zippo Manufacturing Co.

ZIPPO REPLACEMENT AND REPAIR POLICY

1. REPLACEMENT
Any New Zippo Product which fails to work will be replaced Free of Charge when returned to Zippo Manufacturing Co., Bradford, Penna. 16701

2. REPAIR
Any Zippo Product, regardless of age or condition, will be repaired Free of Charge when returned to Zippo Manufacturing Co., Bradford, Penna. 16701

All 1932/33 models had square corners and were approximately 2-7/16 inches tall. The full dimensions were 2-7/16 x 1-1/2 x ½ inches. The top and bottom of the cases were made of seamless tubing, purchased from Chase Bras Co. in Connecticut.

The top and bottom plates of the hinge were soldered flush on the case tubing using a hot plate. The name plate logo on the bottom of the case had both ZIPPO and "PAT. PENDING" stamped on it.

Blaisdell produced both high polish and dull chrome cases early on. Late in 1933 the soldering of the hinge on the case was discontinued and the hinges were then spot welded. Blaisdell used a second-hand Thompson welder at that time. The case and inside unit were plated by John Vitale of Olean, New York.

When assembling the unit, Blaisdell used what he called a "Just Rite" flint wheel. The cases had three-barrel hinges. The inserts were of a seamless, high polished chrome variety, with no patent numbers on the insert. The insert was ¼ inch taller than inserts made today to fill the extra ¼ inch that the bottom of the case had at that time. Zippo actually produced two sizes of inserts in 1932 and 1933. The earliest insert had a "fluid chamber" that was 1-5/8 inches tall. The later insert had a fluid chamber 1-1/2 inches tall.

During 1933 and continuing on through 1936, Zippo made the lighters ¼ inch shorter. The fluid chamber of the insert was, therefore, approximately 1-2/8 inches tall. Around 1934, Zippo designed and began utilizing the "Alanklin Wheel." At the time this model still had the hinge plates on the outside of the case. The "shorter" 1933-35 variety had a three-barrel "outside" hinge, whereas the early 1936 model had a four-barrel "outside" hinge.

Later in 1936 Zippo made one more change: It moved the hinge plates to the inside of the lighter and made the inside unit (insert) a fraction smaller to accommodate it. These smaller units, and all units from 1932 to 1936, had a coiled spring unit to push the cam in place. One relatively easy way to identify this new inside unit was to look under the cam. The coiled spring unit was similar to a ball point pen spring, pushing up a small piece of metal. This coiled spring unit caused tension on the cam, holding it in place and holding the lid shut. The coiled spring unit was located inside a tube within the insert. The tube can be seen if the cotton packing is removed.

The coiled spring/cam unit did not hold its tension very well and needed to be repaired by Zippo after prolonged use. Most surviving early 1936 models have

been repaired by Zippo, using a humped flat spring, which replaced the coil spring unit during 1937. The Zippo repair department would pull out the spring unit and replace it, leaving an empty hole under the cam and the empty tube inside the insert that had held the coiled spring unit. Zippo lighters made from 1932 to 1936 with complete coiled spring units intact are very rare and valuable. They are even more rare and valuable if the spring still has good tension, since most remaining springs have lost their tension.

Between late 1936 and 1941 Zippo lighters, with access numbers 7 through 13 inclusive, had four barrel hinges. Most had "humped" flat spring inserts, except for a few with the early coiled spring variety. During the 1932 to 1941 era, all the inserts were of the seamless tube style, highly polished chrome with no logo or patent numbers on the insert. Around 1938, the first full drawn case was made by Backus Novelty Co. of Smethport, Pennsylvania. The cases all had "moderately" flat bottoms, with 2032695 patent numbers on the bottom of the lighter. The seven variants (access numbers 7, 8, 9, 10, 11, 12, & 13) could be found having line-drawn company trademarks. Sports illustrations, or initials. They could also be found having metallique motifs illustrating company trademarks, early Sports scenes, or a person's initials.

The contrast between a 1932-33 repaired insert (left) with that of a Mid 1933-36 model (right). Note the differences in height.

A Zippo Chronology Milestones in the History of the Great American Lighter

1931 – During the summer of 1931, not 1932 as previously believed, Blaisdell gets the idea to modify and import Austrian lighters.

1932 – Blaisdell develops prototypes and calls his new Austrian lighter, Zippo. Research leads this author to believe that Blaisdell was producing his own "square corner" prototypes as well as "hook cam" tall models towards the end of 1932, which is why I am going to call "hook cam" tall models 1932s.

1933 – Blaisdell starts production of the square-cornered windproof lighter in February. Production models sport two styles. One style is plain whereas the other variant has diagonal lines cut on the face of the lighter. Both of these models are 2 & 7/16 inches tall, ¼ inch taller than the now standard Zippo lighter. The inserts are also approximately ¼ inch taller. Both the case and the insert are shortened the latter part of 1933.

1934 – During 1934 the ¼ inch shorter units continue to be produced with the hinge plates still on the outside of the case. Blaisdell applies for the Zippo patent on May 17, 1934, after test marketing the product for two years.

1935 – Metallique specialty advertising begins with Kendall Oil's order of 500 lighters in 1935.

1936 – Two different models are produced. The first model has a four-barrel hinge instead of a three-barrel hinge with the plates still located on the outside of the case. The second model has a four-barrel hinge as well, but the plates are moved to the inside of the case. The case is still of the square-case variety. Late in 1936 both sterling and engraved specialty advertising models made their debut on the latter variant.

1937 – Sports models enter the market, emblazoned on Zippo's "Classic case." Zippo also introduces gold-filled, solid gold, and engine-turned models according to the 1937 *Esquire* advertisement.

1938 – The brass drawn case with rounded corners awes the market. This model has a soldered clip inside the lid. This style is only produced during 1938 and 1939. Sterling and 14k gold lighters, in both plain and engine-turned styles as well as those with a high polished chromium finish are introduced in this variant.

1939 – Production begins on the 1st model Barcroft table lighter. This model is only produced during 1939 and 1940.

1940 – The brass drawn case with no metal clips inside the lid makes its entrance. It has one piece extended from the hinge that supplants the metal clips. This variation with the "flat bottom" is only produced in 1940 and 1941. Lighters produced today are made the same way with the extended hinge.

1941 – Toward the end of the year Zippo makes its first steel outer case, which retains the four-barrel hinge with a chromed brush finish. The insert is made of steel in lieu of brass.

1942 – Black Crackle models made of steel with a four-barrel hinge are only produced this year.

1943 – Steel Black Crackles with a three barrel hinge are made from 1943 to 1945.

1946 – Zippo fabricates cases made of a nickel-silver alloy between 1946 and 1947. Between 1946 and early 1948, Zippo designs a sterling model with the word "sterling" written under the hinge.

1947 – Zippo produces prototypes and test samples of Town and Country lighters. The "official" Town and Country series takes its place among Zippo standouts in January, 1949 and are produced as late as 1960. The 2nd Model Barcroft table lighter and the "Loop" model or "Loss-proof" become coveted pieces.

1948 – Zippo produces a three-barrel hinge, sterling model, that has the words "Zippo" and "Sterling" written on the bottom of the lighter. The Zippo logo is placed the width and not the length on the bottom of the case.

1949 – The 3rd Model Barcroft table lighter arrives on the scene. It is manufactured between 1949 and 1954. The Lady Bradford table model with no base (hereafter called the 1st Model Lady Bradford) is produced only in 1949, until it was recalled. For the record, Lady Bradfords were not in "official" production as of June 30, 1949, according to a Zippo "net price list." More than likely, Zippo test marketed them for most of the year. January, 1949 marked the beginning of the "official" Town and Country series.

1950 – Zippo manufactures a "Full Cover" Leather model from this year through 1957. The "recalled" 1st Model Lady Bradford, is remanufactured and produced until April 1, 1954. At this time, the base is added to the bottom of the original lighter. Between 1950 and 1954, Zippo produces a sterling model with a five-barrel instead of a three-barrel hinge. The words "Zippo" and "Sterling" travel the length on the bottom of the lighter. Late 1949 "true" 1950 models have a five-barrel hinge in lieu of a three-barrel hinge.

1951 – During the Korean War, Zippo cases were fabricated of steel. Steel is employed until late 1954 when Zippo returns to using chrome plated brass.

1952 – A Wrap-Around Leather model (Zippo model #550) is added to the Zippo line. The 1952 leather-wrap model is made in conjunction with the 1950 full-leather model for a number of years. The leather-wrap model is produced until 1960, whereas the full-leather model is produced until 1957. Zippo uses a wrap-around application for the lid and bottom. These are available in brown alligator or red, brown, and green reptile, as well as in black or blue Moroccan leather.

1953 – The patent number is changed from 2032695 to 2517191, with a large "pat pending" logo. The logo is written in block letters on the insert as well as the bottom of the lighter. Zippo uses this "bottom logo" until mid-1955.

1954 – Zippo begins production of the 4th Model Barcroft table lighter. This model is produced until 1979.

1955 – Zippo introduces both silver and gold-filled lighters with the new bottom logo. Of course, Zippo had been making gold-filled models since 1937, according to Blaisdell's 1937 *Esquire* advertisement. About 1955, Zippo changed the bottom logo on the sterling models. The logo is produced in script, and written horizontally. Zippo begins the use of codes in 1955 on #250 models, and not 1958 as many have suggested. Various combinations of dots and slashes are now stamped on the bottom of all lighters. These marks indicated the year the lighter was manufactured (see "Dating Zippo Lighters"). Zippo applies for a wheel guard patent (2,704,447) for regular size lighters, which they never use. Blaisdell calls it "Flameguard for Flint Wheel." A similar wheel guard variant was issued on 1955-56 slim models.

1956 – Zippo starts production of slim lighters, including gold- and silver-filled models.

1957 – Commercial trademarks using the etching process are first used. The 25th Anniversary Commemorative lighter is produced.

1958 – Zippo begins production of the slim 14k gold lighter. Zippo is awarded the 2517191 patent.

1959 – The logo on the insert is changed from horizontal to vertical.

1960 – Zippo begins production of both the Moderne and Corinthian table models. The Town and Country series comes to an end although a few advertisers are still produced using the Town and Country process through 1964.

1962 – The "Rule" is first manufactured.

1964 – The Pocket Knife with no money clip is first produced.

1965 – The Golf Ball, as is used in the game of golf, makes its debut.

1966 – Belt buckles are added to Zippo's product line.

1967 – The patent number is removed from the bottom of the case.

1969 – In honor of the NASA moon landing, a Moon Landing lighter theme is developed, and becomes highly prized. The theme continues through the years into the 1980s with the latest in the line being the Space Shuttle design. Also, the Greenskeeper is added to Zippo's general line.

1970 – In keeping with the interests of the early 1970s, Zippo introduces the "slim" Zodiac series.

1971 – The Money Clip with Knife is marketed.

1972 – The Woodgrain case, National Football League series, and 40th Anniversary Commemorative lighters make their debut in 1972.

1973 – The Key Holder is first marketed.

1974 – Venetian style cases are produced.

1976 – Three additions were added to Zippo's general line: the pocket Magnifier, the Denim series, and the Bicentennial series. The Denim series is in keeping with the clothing fashion at that time, with a denim-look imprinted with different motifs. The Bicentennial lighter is manufactured in celebration of the nation's 200th birthday.

1977 – Three more additions to Zippo's general line: the Golden Tortoise, Golden Elegance, and Scrimshaw cases.

1978 – Both the Ultralite model and the Letter Opener are first marketed. Both advertising and military logos are printed on the surface of the acrylic chips, as well as Walt Disney logos.

1979 – The Handlite, Gold Electroplate, and Political motif models (Elephant & Donkey) make their premiers.

1980 – The Desk Set is added to the line.

1981 – Both the Pillbox and Cut-About are marketed.

1982 – The 50th Anniversary Commemorative lighter, Pipe lighter, and Patriotic series premiere. Zippo adds writing instruments (pen and pencil) to the line.

1983 – Lighters marked "Solid Brass" on the lid are introduced to the public.

1984 – The black powder coat is introduced. This is applied electrostatically to the brass case. Soon four more colors are introduced: blue, burgundy, green, and grey. A year later the imprinted logo and a border are added. The powder coat produces what is commonly called a "matte finish."

1986 – Camouflage lighters are introduced. The design is surfaced printed in two colors on the matte green powder coat. It is available in regular and slim models. Many powder coat lighters are surface imprinted with logos.

1987 – Zippo adds the Elvis set to the general line. The set includes three regular size lighters as well as four slims.

1988 – Zippo begins manufacturing the 1932 Replicas in 1988. It is similar in height and design to the original 1932/33 tall model, although there are more differences than similarities.

1990 – Five series sets were added to Zippo's line: the Wild West series, 1st Anheuser-Busch series, Presidential series, Fabulous 50's series, and Super 60s series. The Knife/Scissors combination makes its debut.

1991 – The Civil War series, American Classics series, Wildlife series, and Peterbilt Truck series are all added to the general line.

1992 – Three lighters are introduced: the 60th Anniversary Commemorative lighter, the 60th Anniversary Commemorative Set, and a sterling with gold inlay, anniversary commemorative version. The 60th Anniversary Commemorative lighter is introduced in a collectible tin. Zippo also produces a 60th Anniversary lighter set that contains replicas of the 5th, 10th, 25th, 40th, 50th, and 60th commemorative years. Also, as a way of saying "thanks for a job well done," Zippo produces approximately 1200 sterling units with gold inlay for its employees. Lighters exhibiting a midnight chrome finish are added to Zippo's product line.

1993 – Zippo buys Case Cutlery and begins production of lighter/knife combination sets. The Born to Ride series, Geometric series, Motorsports Collections, Varga Girl, Vintage Aircraft, and Corvette series premiere.

1994 – Zippo adds many lighters to its general line: the D-Day Commemorative lighter, Allied Heroes set, 2nd Anheuser-Busch series, Barrett-Smythe Collection series sets (Backyard Insects, Animal Friends, Dinosaurs, Endangered Animals, Toledo, & Wildlife), Sport and Game series, Souvenir Truck series, Stargate, Jeff Gordon, Buffalo Bills, Woodstock, Christmas set, and L'Art Classique set.

1995 – Zippo produces these sets: Valentine, Mysteries of the Forest, Antique Finish, Marble, Jim Beam, Chevrolet, Potpourri, Track, Driver, Car, Event, Racing Trademark, Centennial Olympic, World War II, 3rd Anheuser-Busch series, Barrett-Smythe, Barrett-Smythe Emblem, and Barrett-Smythe Comic Strip sets. Zippo also produces the 18k gold lighter instead of the 14k model in both regular and slim models.

1996 – Zippo introduces many new sets: Pinup Girls, Barrett-Smythe Comic Strip, World War II Remembrance, Chevrolet, Ford, James Bond Movie Posters, James Bond Golden-eye, Ted Lapidus, Driver, Event, Track, Racing Trademark, Dale Earnhardt, Car, 4th Anheuser-Busch series, Corona, Red Dog, Colt Firearms, Camel, Miller, Jim Beam, Barrett-Smythe Trick Lighter, Cigar Store Indian Lighter, Toledo Black Matte, Olympic Games Centennial, Southwest Collection, Midnight Chrome, and Pewter Wild West sets.

The Lighters

*Concerning Pricing the Rarest & Most Desirable Zippo Lighters

I personally know of four 1932/1933 tall models that have the old "hook style" cam that sold in the $30,000 range about five years ago. Today, I don't believe that you could buy one for even twice that amount, unless the seller didn't realize what he had. Therefore, it is near impossible to put a price on such items. Based on this premise, items that fall into this category, will be listed as "**too rare to value**." I believe that items of this type, in the near future, should and will be sold at auction houses such as Sothbys and Christie's. The rarest and most desirable items, in most true collectible fields, dwarf the prices paid for a 1932/1933 tall model five years ago.

My next book, is scheduled to go to the publisher yet this year. It will include not only the rarest Zippo lighters but also the rarest boxes. In many cases, early boxes are worth considerably more than the lighters that go in them. I believe this is true of the 1932/1933 tall model box as well as many of the other early boxes.

I have given every lighter an access number. I added letters of the alphabet to the access number to show the different variations of the basic lighter. Also, I have written a price range by the picture itself. **(Note that all lighters are valued in excellent condition regardless of the condition in the photograph unless otherwise noted.)** Charts and a more detailed description of how I arrived at those prices are in the back of the book. All lighters are regular size unless stated or illustrated differently.

The First Models: 1932-1943

Access numbers 1-8 have square corners producing 90 degree angles. They also have soldered cases.

U-shaped cam stop on Access numbers 1-10.

Access numbers 1-6 have eight holes located on each side of the chimney. Access numbers 7-14, 27, 33A, and 33B have seven holes on each side of the insert's chimney, which Zippo started doing circa 1936. Zippo again started putting eight holes on each side of the insert's chimney beginning circa 1947, and is still doing this today.

Access numbers 1-10 have the U-shaped cam stop soldered in the inside of the lid.

Access numbers 9-12 are the brass drawn case variety, with rounded corners. The inserts in numbers 9-12, are brass drawn tube style. This insert was also used in 1937. The cases have a flat bottom that is slightly pushed outward. In contrast, access numbers 1-8 had totally flat bottoms and sharp 90 degree angle corners.

#1: 1932/33 Tall Model (No diagonal lines on case)

This is the first model that Blaisdell produced. This model is approximately ¼ inch taller than models manufactured today. The case is 2-7/16 inches tall, also making it ¼ inch taller than 1934-36 outside hinge plate models. "Hook cam" 1932/33 Tall models are worth considerably more than "figure eight cam" models taht were produced during the same time frame.

#1 A – Plain
    #1 A1 – Solid Brass prototype (neither the insert nor the case was ever chromed)
#1 B – Metallique Initials
#1 C – This variety has a generic metallique applied to it.
#1 D – This lighter has the original hinge but the piston, in the insert, has been removed by Zippo and replaced with a humped or flat spring.
#1 E – This lighter does not have the original hinge on the case or the original piston in the insert. The hinge, piston, cam, and flint wheel have been repaired and replaced by Zippo. The piston is replaced with a flat or humped spring.
#1 F – Very early 1932/33 Early Hook Cam Brass prototype. Note the hook cam and shorter body.

Circa 1937-1941 insert.

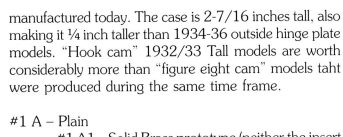

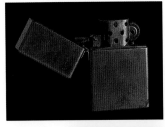

#1F, Front, opened.

#1F, Front, closed.

#1 F – 1932/33 Brass Prototype
Too rare to value.

#1 A, hinge view.

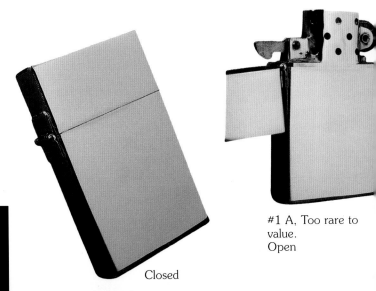

#1 A, Too rare to value.
Open

Closed

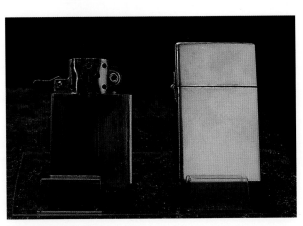

#1 A, disassembled.

18

#1 A, close up of the cam and piston.

#1 A, piston and raised hook cam view

#1A1, closed. Too rare to value.

#1A1, open

#1 E, closed, $4,000-$5,550

#1 E, open

#1 E, with replaced wheel

1932-35 models, that have the original hinge must have a three barrel hinge similar to the lighter on the right whereas hinges that have been replaced by Zippo usually have four or five barrel hinges similar to the lighter on the left. Note the cleat marks that are found above and below the hinge plates on an original model. Original models must have these cleat marks, repaired lighters don't, due to the hinge plates covering up the cleat marks on a repaired lighter.

#2 – 1932/1933 Tall Model (Has diagonal lines on case)

This model is still ¼ inch taller but has diagonal lines cut on the face of the lighter.

This illustration contrasts the repaired hinge of a 1932/33 on the left, that was done by Zippo, with an original on the right. "Hook cam" 1932/33 Tall models are worth considerably more than "figure eight cam" models, that were produced during the same time frame.

#2 A – Plain

#2 A1 – This variant has all its parts except the cam & washers, which were repaired and replaced by Zippo. Zippo left the original coiled cam spring intact. It is extremely rare to find this model repaired in this way.

#2 B – Metallique Initials.

#2 C – This variant has a generic metallique applied to it.

#2 D – This lighter has the original hinge but the piston & cam, in the insert, has been removed by Zippo and replaced with a humped or flat spring.

#2 E – This lighter does not have the original hinge on the case or the original piston in the insert. The hinge, piston, cam and flint wheel have been repaired and replaced by Zippo. The piston is replaced with a flat or humped spring.

#2 A, closed, too rare to value.

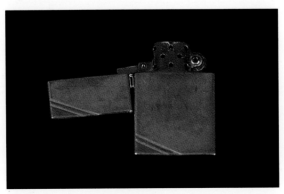

#2 A, open

#2 A 1

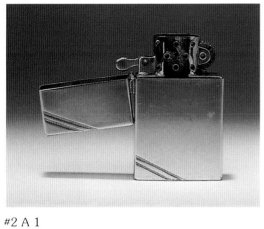

#2 A 1

#2 A 1

#2 B, Too rare to value (if not repaired)

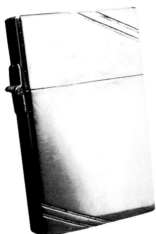

#2 A 1

#2 D, $6,000-$7,500

#2 E, plain, $4,500-$5,500

Box for #1A & #2 A
The box is worth as much or more than the lighter that goes in it. The Zippo Museum doesn't own a 1932/33 box. Too rare to value.

#4 A, $2,800-$5,000

#4 B, $2,800-$5,000

#5 1, obverse, $7,000-$10,000 (if totally original)

#5 1, reverse, $7,000-$10,000

#3 – Mid 1933 Short Model

1933 models without lines. During this time frame the case was shortened ¼ inch. The hinge plates were still located on the outside of the case and still had three barrels attached to them. A few 1933s were produced that were shorter, like those produced later on in 1933, 1934, and 1935, but they didn't have diagonal lines cut on the face of the lighter. They are extremely rare.

#3 A – Plain or has engraved initials
#3 B – Metallique Initials
#3 C – This variant has generic metallique advertising applied to it.

#4 – Late 1933-35 model (with diagonal lines)

This model is similar in design to access #3, with the exception that it has diagonal lines.

#4 A – Plain
#4 B – Metallique Initials

#5 – 1935 model

1935 marked the beginning of "Specialty Advertising" on lighters. It is impossible to tell the difference between a plain late model 1933, a 1934, or a 1935 model. During this time frame some advertising logos were done using metalliques and some were line drawn. They were all manufactured with three barrel hinges and the patent pending logo was still stamped on the bottom of the case.

I placed all advertising motifs in access #5 since 1935 marked the beginning of "specialty advertising" for Zippo.

#5 A – Oil Company Engraved Advertisers
#5 B – All Other Engraved Generic Advertisers
#5 C – Kendall Metallique
#5 D – Drunk Metallique (circa 1935-39)

5D & 5E

#5 D is actually a line drawn illustration but it provides you some idea of the metallique motif. #5 D (Circular Border): #5 E (Rectangular Border). Both have the same basic design.

#5 E – Drunk Metallique (circa 1935-39)
#5 F – Scotty Dog Group Metallique
#5 G – Gulf Metallique
#5 H – I grouped all other generic metallique advertisers together as far as price and rarity.
#5 I – Salesman's Sample (Includes Town and Country Sloop, Sports Fisherman, and Flag Emblem motifs)

#6 – 1936 Outside Hinge model

This model still has the hinge plates on the outside but it is different than the preceding models in that is has four barrels in lieu of three barrels attached to the hinge plates. This model may or may not be found having advertising on it. This was the last model to have "patent pending" stamped on the bottom of the case.

#6 A – Plain
#6 B – Metallique Initials
#6 C – Engraved Generic Advertising
#6 D – Kendall Metallique
#6 E – Scotty Group Metallique
#6 F – Drunk Metallique (front view)
#6 G – Drunk Metallique (side view)
#6 H – Gulf Metallique
#6 I – I grouped all other generic metallique advertisers together as far as price and rarity.
#6 J – Double Tank Prototype

#6 J, $12,000-$16,000

#6 J, disassembled

It has been stated, in different literature, that the square models with the hinge plates on the inside were made until 1940. Much has been documented. It is this author's opinion, based on original Zippo salesman's catalogs, leaflets, and advertising brochures, that standard production lasted until 1940. Whether they were made until 1937 or 1940 is unimportant to the collector who is trying to put together a world-class collection. What is important is whether one has collected all the different variations from 1932-42, as well as those to date. This author will call these 1936-40 models, with access numbers 7 and 8.

Zippo produced the 1936-40 square case models in conjunction with the brass drawn cases until 1940, but Zippo continued to produce the brass drawn cases into 1941. Remember, there were two styles of brass drawn cases. Those with the U-shaped soldered cam stop (1938-39, access numbers 9 & 10) and those with the extended hinge cam stop (1940-41, access numbers 11 & 12). An extension of the hinge which went up into the lid and curled around, supplanted the function of the previously soldered cam stop. Zippo used these two almost completely different methods of production to fabricate the two models. The flat top and bottom of the square case model (access number 7 & 8), needed to be soldered on, whereas the 1937-41 brass drawn case models (access numbers 9-12), didn't. Both the 1936-40 square case and the 1938-39 brass drawn case models, needed to be soldered where the U-shaped cam stop attached to the lid. I would have thought that Blaisdell would have made the transition in production solely to the brass drawn case to eliminate much of the soldering process. Brass drawn cases that eliminated the soldering process would have increased efficiency, productivity, and profit. The character, style, and desirability of the square case model, more than likely, prevented this from happening. Being the entrepreneur that Blaisdell was, he would have tried to make the transition in production from the square to the brass drawn case as soon as possible. The advent of World War II brought and end to the "classic" square cased model as well as the brass drawn case model.

Access numbers 7 and 8 are of the soldered case variety that have square style corners producing 90 degree angles. Both have totally flat bottoms, with the 2032695 patent on the bottom of the case. The 1936 model is the first model to have the 2032695 patent. All the preceding models had "patent pending" on the bottom of the case. Both have hinge plates that are located on the inside of the case, with only the hinge barrel sticking out. Access numbers 7 and 8 can be found having either a piston spring insert or a humped-spring insert. Early examples have a piston spring insert, whereas later models have a humped leaf spring insert that applies tension to the cam to hold it in place.

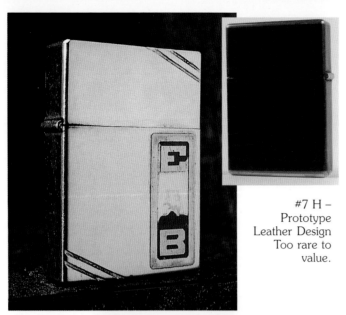

#7 H – Prototype Leather Design Too rare to value.

#8 B, $1,000-$1,500 The illustration is missing a metallique initial.

#7 A1 - $7,500-$10,000

#7 F – 1937 Gold-Filled (plain) in Esquire Ad
#7 G – 1937 Gold-Filled (engine-turned) in Esquire Ad
#7 H – Prototype Leather Design (done in enamel, not leather)

#8 – 1936-40 model

This model has diagonal lines cut on the face of the lighter.
#8 A – Plain
    #8 A1 – With original piston insert and strong spring
    #8 A2 – With original piston insert but weak spring
    #8 A3 – Those with "original humped" spring inserts
    #8 A4 – Piston in the insert has been removed by Zippo and replaced
#8 B – Metallique Initials
#8 C – Kendall Oil Metallique
#8 D – Engraved Generic Advertising
#8 E – Generic Metallique Advertising
#8 F – Drunk Metallique

#8 F – Drunk Metallique $5,000-$7,000

U-shaped Cam Stop

#8 C, $7,500-$10,000

#7 – 1936-40 model (Prototype design)

This model doesn't have diagonal lines cut on the face of the lighter. This is an extremely rare case. I have seen very few examples. It is a non-production prototype design. This variant has square cut corners and is most often found in a nickel/silver finish. Access numbers 7A1-7A4 are variations of this model.
#7 A – Plain
    #7 A1 – With original piston insert and strong spring
    #7 A2 – With original piston insert but weak spring
    #7 A3 – Those with "original humped" spring inserts
    #7 A4 – Piston in the insert has been removed by Zippo and replaced
#7 B – Engine-Turned (were first manufactured in 1936)
#7 C – Sterling "with square corners"
#7 D – 14k Gold "Plain"
#7 E – 14k Engine-Turned

#9 – 1938-39 model

This model has no diagonal lines cut on the face of the lighter. This variety has a U-shaped piece of metal that is soldered inside the lid. The U-shaped soldered "clips" act as a cam stop that engages the lid to hold it down and the lighter shut.
#9 A – Plain
#9 B – Metallique Initials
#9 C – Engine-Turned in "Chromed" Brass
#9 D – Plain Sterling (may or may not have initials)
#9 E – Engine-turned in Sterling (Note that the middle hinge barrel is longer on both #9D and #9E sterling models.)

#9 F – Engraved Generic Advertising
#9 G – Generic Metallique Advertising
#9 H – 14k Gold Model
#9 I – Belle Kogan Series Designs (Zippo model numbers K1, K2, K5, & K7; with black enamel inlay; regular production models)
*See 1938 Esquire advertisement on page 177 for description.
  #9 1 – K1
  #9 1 – K2
  #9 1 – K5
  #9 1 – K7
#9 1-K7 and 9 1-K5 are two Belle Kogan Designs that were brass with gold plate. They were never chrome plated with enamel inlay. These are prototype examples and are far rarer and worth more than regular production pieces.
#9 J – High Polished Chrome Case
#9 K – Gold Filled (plain) Example in Esquire ad
#9 L – Gold-Filled (engine-turned) Example in Esquire ad
#8 M – Sports Motifs and Facsimile Signature (Extremely rare to find three different engravings as well as being highly polished on this model)
#9 N – Sailfish (Access #17 E) Variation
#9 O – 1938 Slim Test Model (Zippo's first attempt at a slim lighter. Note the small three barrel hinge and how Zippo's R&D department took a regular size model and cut it down. This also has a small U-shaped cam stop.)
#9 P - Test model insert.

#9 A, $650-$1,000

#9 C

#9 E, $5,000+   #9 I-K2, obverse, $3,500-$5,000   #9 I-K2, reverse

25

Both 9I-K5 and 9I-K7 are prototypes that were originally gold plated. The plating was so thin that it literally vaporized over the years. Both are new, never lit.

#9 I-K5, $10,000-$15,000

#9 I-K7, $10,000-$15,000

#9 M, reverse

#9 M, obverse, $3,500-$4,500

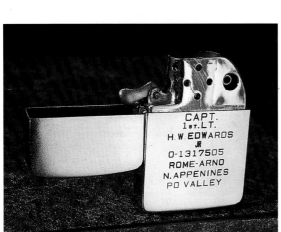

#9 P, only one known to exist. Too rare to value.

Bottom four are all #9 O. Too rare to value.

#10 – 1938-39 model

This model has diagonal lines cut on the face of the lighter as well as having the same soldered U-shaped cam stop inside the lid, as access #9.

#10 A – Plain
#10 B – Metallique Initials
#10 C – Engraved Generic Advertising
    #10 C1 – HYP
#10 D – Generic Metallique Advertising
#10 E – High Polish Chrome (extremely rare)
#10 F – Metal Emblems (put on by the owner)

I don't cover the 1939 1st Model Table lighter at this time. I cover it at a later point with the other Table Models. (*See access #157*)

#10 C1. $850-$1,250

#10A, $600-$1,000

#10 D, $1,000-$1,500

#9J, $850-$1,250

#11 C1, $850-$1,250

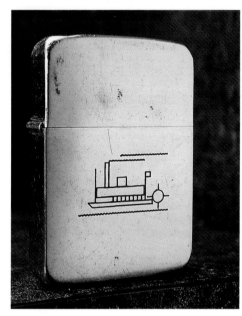

#11 C2, $850-$1,250

#11 C3, obverse, $1,000-$1,500. This lighter is double-sided which makes it worth more.

#11 – 1940-41 model

This model has no diagonal lines cut on the face of the lighter. It is different from the previous two models in that it lacks the two soldered clips. The cam stop is actually an extension of the hinge that extended into the lid and curled back around to supplant the U-shaped clips. This is the first model that Zippo produced in this way. Zippo still produces the cam stop in this way.

#11 A – Plain
#11 B – Metallique Initials
#11 C – Engraved Generic Advertising
 #11 C1 – National Products Advertiser
 #11 C2 – Delta Queen
 #11 C3 – Elastic Stop Nuts (Rare to find this model with advertising on both sides)
 #11 C4 – American Roller Bearing
 #11 C5 – KEROTEST
 #11 C6 – YALE
 #11 C7 – INSELBRIC
 #11 C8 – ZIMMER SPLINT COMPANY
 #11 C9 – FRASSE STEELS
#11 D – Generic Metallique Advertising
#11 E – Sterling Silver Model (may or may not have initial plate)
#11 F – 14k Gold Model
#11 G – Belle Kogan Series Designs (Zippo model #'s K1, K2, K5, & K7; with black enamel inlay) *See 1938 Esquire Advertisement on page 177 for Description.
 #11 G – K1
 #11 G – K2
 #11 G – K5
 #11 G – K7
#11 H – High Polished Chrome
#11 I – Gold Filled (plain) in *Esquire* ad
#11 J – Gold-Filled (engine-turned) in *Esquire* ad
#11 K – Town and Country Lily Pond Scene with R.F. Thiele's name engraved on the front of the case. (This is probably one of Zippo's first attempts at producing this illustration, which eventually became part of the Town and Country series. The motif was air brushed when Zippo added it to its Town and Country series. This is extremely rare, to say the least. Notice the deep etching, which gives the appearance of high relief.)
#11 L – Elsie Borden
#11 M – Roller Skater (line drawn sports test model)

#11 C3, reverse

#11 C6, $850-$1,250

#11 C4, $850-$1,250

#11 C7, $850-$1,250

#11 C5, $850-$1,250

#11 C8, $850-$1,250

#11 C9, $850-$1,250

#11 E, $5,000+. This 1940s sterling model belonged to Bill Edwards, one of Blaisdell's first employees.

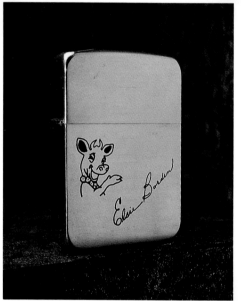

#11 M, $1,000-$1,500

#11 K, $3,500-$5,000 (if in excellent to mint condition)

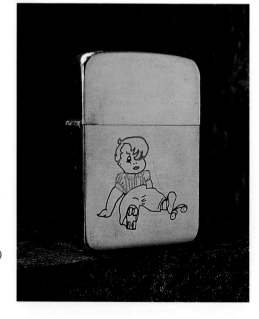

#11 N, $4,500-$6,000

#12 – 1940-41 model

This model has diagonal lines cut on the face of the lighter as well as having the same extended hinge/cam stop, like access #11.

"Reverse Engraved" prototypes and test models were produced by Zippo in the late 1930s, early 1940s. In this process the illustration is produced when the background of the lighter's surface is removed (cut away) by etching. It is similar to the negative of a photograph. *See access numbers 12 B and 12 C.* The technology of the time prevented Zippo from "mass producing" the illustrations.

#12 A – Plain
#12 B – Lakes to Sea Stages (reverse engraved prototype advertiser)
#12 C – Wabash Valley (reverse engraved prototype advertiser)
#12 D – Metallique Initials
#12 E – Engraved Generic Advertising
    #12 E1 – Loaders
    #12 E2 – Instrument Design Inc.
    #12 E3 – Hastings
    #12 E4 – H-H-Y-C
    #12 E5 – "Happy Days"
    #12 E6 – LBR. CO.
    #12 E7 – U.S. AIR CORPS
    #12 E8 – DRAKE
    #12 E9 – "HARTER" STEEL CHAIRS
#12 F – Generic Metallique Advertising
#12 G – 14k Gold Model
#12 H – Military Emblem (Zippo applied)
#12 I – Engraved Crossed Guns
#12 J – "Trench Art" Emblems (Soldier applied)
#12 K – Multi-scene Zippo Employee Test Model

Around 1941, due to the advent of World War II, cases and inserts were fabricated out of chromed steel. Circa 1942 Zippo used a plain steel case with a black crackle finish in lieu of a chromed finish due to the shortage of chrome. The black crackle finish was both masculine for the military and aesthetically attractive. It prevented the steel case from rusting but wasn't very durable and the original paint wore off quite easily. Many soldiers would then decorate their World War II models with elaborate engraved illustrations, engraved sayings, and "trench art" medallions, which often made a statement concerning the soldiers' identity and values.

#12 A, $650-$1,000

#12 B, $1,500-$2,000

#12 C, $1,500-$2,000

#12 K, Front and back. Multi-scene 1940-41 Zippo Employee Test Model. Unbelievable! Too rare to value.

#12 E, $850-$1,250

#12 E, $850-$1,250

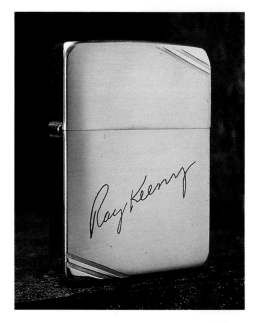

#12 E, $850-$1,250

#12 E, $850-$1,250

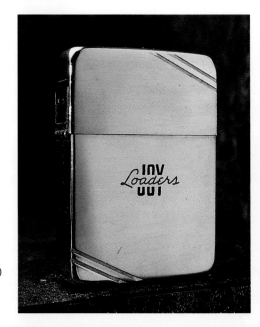

#12 E1, $850-$1,250

#12 E2, $850-$1,250

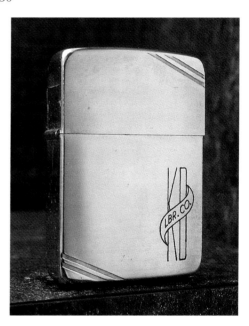

#12 E5, $850-$1,250

#12 E3, $850-$1,250

#12 E6, $850-$1,250

#12 E4, $850-$1,250

#12 E7, $850-$1,250

#12 E8, $850-$1,250

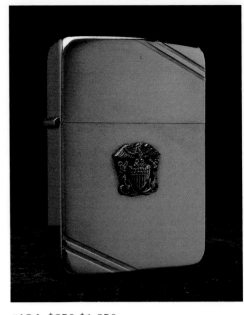

#12 I, $850-$1,250

#12 E9, $850-$1,250

#12 J, reverse, $650-$850 (if in excellent to mint condition)

#12 J, obverse

#12 I, $850-$1,250

#12 J, $650-$850

#13 – 1941 Steel Drawn Case model

Late in 1941, Zippo made a steel cased lighter with a four barrel hinge and a brushed finish. This model was produced in the same way as numbers 11 and 12, with exception that the case is made of steel and can be identified using a magnet. The insert for this model was first fabricated in brass and then steel. Steel inserts, for this model, are extremely rare. Both inserts, for this model, are identical in design to those fabricated circa 1937-40. Therefore there is no seam, in the fluid chamber, under the cam.

#13 A – Plain
#13 B – Engraved Generic Advertising
    #13 B1 – "Lion Logo"

#14 – 1942 BLACK CRACKLE

The 1942 black crackle is known as the first World War II model. This was also the first year of the black crackle finish. The 1942 model was similar in design to the 1940-41 model with the exception that it was made of steel, instead of drawn brass. The insert, as well as the case, was made of steel and not brass. This model had a four barrel hinge similar to previous models produced from 1936-41. The steel composing the insert was wrapped opposite that of other inserts. The insert originally had luster. The 1942 model doesn't have any Zippo logo on the insert. The flint wheel has horizontal cuts on it. The chimney has seven holes on each side with sharp vertical sides.

#13 B1, $850-$1,250. Logo says "God is my Right"

In 1942 Zippo produced an irregular model with a #203695 patent number. This patent number lacks the number 2 between the numbers 3 and 6. It was only produced in 1942. The black crackle paint has to wear off on the bottom of the lighter to identify this model. These 1942 models can be found having military insignia, engraved initials, and engraved Sports motifs, which were produced by Zippo.

#14 A – Plain or with Engraved Initials
#14 B – Engraved facsimile signature done by Zippo
#14 C – Irregular model, with pat. #203695 (missing the #2 in pat. number)
#14 D – Metal Emblems (insignias) done by Zippo
#14 E – Pipe Smoking Fishermen (see #17 I)
#14 F – "Trench Art" Military Emblems
#14 G – Engraved Crossed Guns
#14 H – Engraved Generic Advertising
    #14 H1 – Curtiss/Wright Logo
    #14 H2 – "Bubbles" Logo (This lighter belonged to "Bubbles" LaRue Hutton, the niece of Dale Hutton who was a long time Zippo employee. Note that the word "Bubbles" has gold inlay.)

These prices reflect lighters in excellent to mint condition, regardless of the condition of the lighter in pictures.

#14 D, $500-$700

$14 H1, $700-$900

#14 D, $500-$700

#14 H2, $1,200-$1,500. This lighter is in mint condition and the word "Bubbles" has gold inlay. I am not quite sure how Zippo made this prototype.

The Sports Series

Zippo started making the Sports series circa 1937 and the series is still in production today. Sports models produced between 1937 and 1941 can be found having both metallique and line engraved drawings. Metalliques, in the Sports series, are extremely rare. I have seen a few 1942s, access number 12, that are line engraved with Sports illustrations. I have never seen nor heard of a lighter manufactured after 1938 that had a Sports metallique emblem on it, but some could exist. There seem always to be exceptions to the rule.

Many of the line engraved lighters that were made from 1939-1960 overlapped, having exactly the same design without regard to yearly codes. This is also true of the color filled lighters that were made between 1951 and 1960. It is this author's feeling that, within reason, the pictures themselves make the lighter desirable and rare, without regard to year codes, and not taking condition into consideration. Those that have both the 2032695 patent and three barrel hinge command a higher value. Also the 1939-42 models that are line engraved, command substantially higher prices than their 1946-60 counterparts.

The Sports Series can be broken down into eight distinct periods of time based on the type of illustration and engraving process used.

Almost from the inception of the Sports Series, Zippo produced many animals that were considered part of the series. Included were various examples of wildlife such as bears, dogs, and deer. One Zippo advertising brochure, found in Zippo's archives, illustrates two styles of deer on a Sports advertisement. Zippo considers the Game Series that is produced today to be an extension of the Sports Series, although it maintains its autonomy. Likewise the Town and Country Series maintained its autonomy, although it also was an offshoot of the Sports Series. Zippo grouped many of the illustrations produced during the 1940s and 1950s under the banner "Sports Series," while using many different engraving and coloring techniques.

Zippo also used Sports motifs on advertising lighters if the illustration met the customer's need. Zippo did this with many Town and Country motifs and Town and Country test models as well. Zippo would use test model illustrations, which were kept in its files, as advertising logos for various companies. An example of this is found on page 75. When making a careful observation of the different illustrations, one can see the natural transition of the different motifs within a series, as well as the transition from one series to the next. One can see the natural progression from the line drawn Sports Series to the Town and Country "Sports" Series without much difficulty. One can also easily see the transition from Town and Country "Sports" motifs to silk-screened Sports illustrations.

Customers could send their lighters back to Zippo after many years of use to have them customized. It is possible to find a Sports scene (line drawn) on a 1932/1933 tall model if it had been sent in with the $1.00 for a change. Also, for $.50 to $.75, customers could have had the metallique of a Drunk, Scotty Dog group, metallique Shooter, etc., put on a 1932/1933 model, or any subsequent lighter, until that service ended. During the same time frame, for $1.00 customers could have had metallique initials applied to the surface in three styles. Additionally, if someone sent his or her lighter back to Zippo wanting to have a "Sports" metallique or the line engraved drawing of a "Sports" illustration put on it at any time, Zippo would have probably filled the order, if possible, not wanting to lose the sale and more importantly, a satisfied customer. Zippo, then and now, was more interested in sales and the satisfied customer, than in setting collecting boundaries for certain design changes at certain years. This is one reason why line drawn illustrations have appeared on Zippo cases as late as 1962. Therefore don't be concerned if you find an older lighter with a more recent Sports illustration on it. Note access #5I. This service ended in the early 1960s. The cost at that time was still $1.00.

PERIOD 1

Period 1 could also be called the "Metallique Period" which started in 1937.

#15 – 1937-1940 Sports Metalliques

The case for this illustration has "square" corners, like an outside hinge, but the hinge plates are on the inside of the case (access #'s 7 & 8). Only two examples are known to exist: that of a boy fishing on a dock and that of a shooter. However, I am convinced that Zippo produced other Sports illustrations on this model.
#15 A – Boy on Dock
#15 B – Shooter

#16 – 1938-1939 model

This model is of the type that has access numbers 9-10. Only a couple of examples are known to exist: the metallique shooter and elephant.
#16 A – Shooter
#16 B – Elephant

PERIOD 2

Period 2 could also be called the "Line Drawn Period."

#17 – 1939-1951 Line Drawings

Simple line drawing motifs were produced between 1939 and 1951. Some variations of line drawn illustrations were produced as late as 1962. The front of this lighter has the 1962 Hunter motif on the front and the line drawn Shooter on the back.

The front has a period five illustration of the Hunter motif.

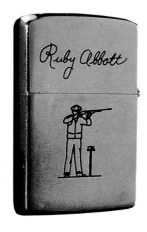

The back of the same lighter has the line drawn period two Shooter.

Most line drawn illustrations were made from 1939-1951. It has been said in some literature that the pipe smoking fisherman (175 E), downhill skier with cape (175 F), male bowler (175 G), baseball (175 J), and football (175 K) were all discontinued by 1950, with access numbers 17 J, 17 K, 17 O, 17 T, and 17 V. I believe, for the most part, that this is true, but there have been some examples found after 1950, although few are known to exist. Customers had a choice of 19 basic line drawn Sports motifs between 1939 to 1960, according to catalogs, leaflets, etc.

These line drawn examples were officially made by Zippo between 1939 and 1951: 175 A, golfer; 175 B, dog; 175 C, sailfish; 175 D, skeet shooter; 175 E, fisherman; 175 F, skier; 175 G, bowler; 175 H, horse head; 175 I, elephant; 175 J, baseball (1940-42); 175 J, sailboat (1946-51); 175 K, football (1940-42); 175 K, crossed bowling pins (1946-51); 175 L, hockey player; 175 M, basketball player; 175 N, baseball player; 175 O, tennis racket; 175 P, bucking bronco; 175 Q, woman bowler.

Both the baseball and sailboat had the same model numbers (175 J) as did the football and crossed bowling pins (175 K), due to appearing on different brochures at different times.

In addition to the 19 basic illustrations, Zippo made line drawn "variations" of many of the 19 motifs that were produced from 1939 until as late as 1960. (I have even seen a line drawn shooter on the reverse side of a 1962 hunter.) I have tried to document and illustrate all the different variations. The variations of each illustration have the same model numbers as the basic design. For example, there are four variations of the downhill skier, which Zippo gave the 175 F illustration number. These have access numbers 17 K, 17 L, 17 M, and 17 M1.

Most people would not think of 1942 World War II models as having line drawn motifs. I have a 1942 "Pipe Smoking Fisherman" in my own collection. Therefore, I have to acknowledge the possible existence of other line drawn sports models on 1942s. Price ranges for Period 2 lighters are for three-barrel hinge models, regardless of the lighter in the photo. 1930s & 40s models pictured here are significantly higher.

#17 A – Golfer (model #175 A)

#17 B – Golfer – with square head (model #175 A)
#17 C – Scottish golfer – with hat (model #175 A)

#17 D – Dog (model 175 B)

#17 E – Sailfish – with single lines (model #175 C)
17 E1 – 1949-50 Sailfish (prototype). This model was never produced in high polish chrome for regular production. Note the slight design change as well.

#17 F – 1954-55 Sailfish – with double lines (model #175 C)

#17 G – Shooter – <u>with cap</u> (model #175 D)
#17 H – Shooter (model #175 D)

#17 I – Fisherman – with a pipe (model #175 E)

#17 J – Fisherman – bending over (model #175 E)

#17 K – Skier – with scarf (model #175 F)

#17 L – Skier – with no scarf (model #175 F)

#17 M – Skier – with front view (model #175 F)
    #17 M1 – Skier – slight variation of #17 M (model #175 F)

#17 N – Bowler (model #175 G)
#17 O – Bowler (model # 175 G)

#17 P – Bowler – with line under him (model #175 G)

#17 Q – Horse (model #175 H)

#17 R – Elephant (model #175 I)

#17 S – Sloop (model #175 J)

#17 T – Baseball (model #175 J)

#17 U – Bowling ball & Two Pins (model #175 K)

#17 V – Football (model #175 K)

#17 W – Hockey (model #175 L)

#17 X – Basketball Player (model #175 M)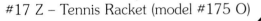

#17 Y – Baseball Player (model 175 N)
#17 Z – Tennis Racket (model #175 O)
#17 ZA – Bucking Bronco (model #175 P)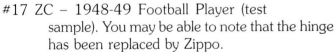
#17 ZB – Woman Bowler (model #175 Q)

#17 ZC – 1948-49 Football Player (test sample). You may be able to note that the hinge has been replaced by Zippo.
#17 ZD – Dog (possible test sample-model #175 B)
#17 ZE – 1938-39 Bear (test sample). This motif is found on access #9M.
#17 ZF – 1950s Canadian Curler
#17 ZG – 1948-49 Prototype Design (has numerous illustrations on both the front and back of the lighter, including miniaturized illustration of 17 H)
#17 ZH – Test model Roller Skater (The design for this motif can be found at access #11M)

Factoid: There were 17 line drawn motifs on the 1949 "Sales List."

#17 A, $800-$1,200. Worth considerably more since it is a 1940-41 model.

#17 A, $350-$500

#17 ZA, $350-$500

#17 ZC, $1,000-$1,500

#17 C
This Scottish golfer is found on a 1938-39 sterling model. Extremely rare and "too rare to value."

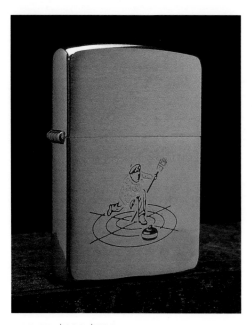

#17 ZF, $350-$500

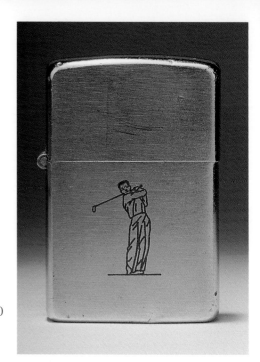

#17 B, $350-$500

#17 ZG, reverse

#17 ZG, obverse, $3,500-$5,000

#17 E, $850-$1,250. Worth considerably more since this is a 1940-41 model.

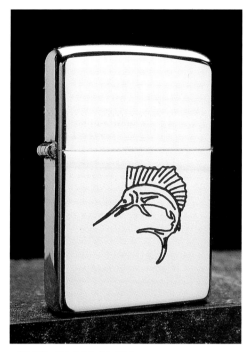

#17 E1, $1,000-$1,500. Note: Prototype sports models were often produced on "high polish" lighters. It is my feeling that high polish chrome accentuates the illustration.

#17 G, $350-$500

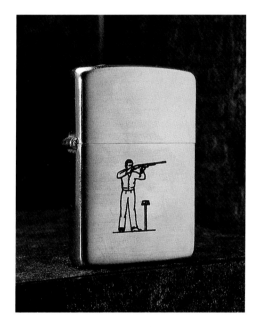

#17 F, $350-$500

#17 H, $350-$500

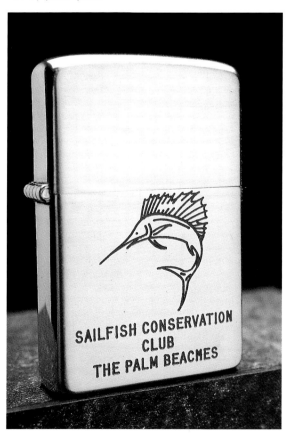

#17 I, obverse, $800+. This particular example is illustrated on a WWII, 1942 steel model. It is extremely rare and worth $800 in this condition, since it has the line drawn fisherman on the reverse side.

#17 I, reverse. $800+

#17 J, $350-$500

#17 K, $350-$500

#17 J, $350-$500

#17 M1, $350-$500

These lighters have to be in excellent to mint condition to be worth these prices, unless otherwise noted.

#17 N, obverse, $350-$500

#17 N, reverse

#17 O, $350-$500

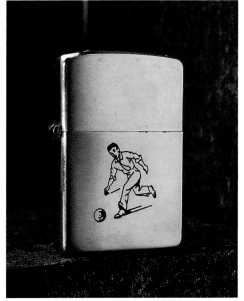

#17 N, $350-$500

#17 P, $350-$500

#17 Q, $850-$1,250. Worth considerably more since it is a 1940-41 model. If not, it would be in the same price range as the other illustrations.

#17 Y, $350-$500

#17 U, $350-$500

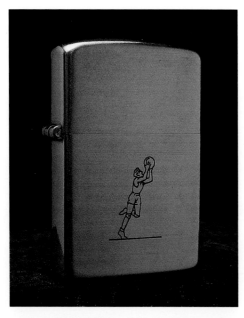

#17 X, $350-$500

During 1951, line design changes and new methods of engraving and coloring were adopted, due partly to the Korean War and the change to steel cases. Also during this same time period, colored illustrations with no background engravings were added to the series. Many of the lighters made from 1951-1959 overlapped, having colored graphics with the same design on them, without regard to yearly codes. Again it is this author's feeling, from the prices he has seen, that the pictures themselves are what makes the lighter desirable and rare, without regard to year codes, within reason, not taking condition into consideration. Zippo started putting color engraved drawings in the background in 1959. Between 1959 and 1970 graphics were engraved in the cap as well as the bottom of the case of the lighter. Zippo no longer engraves Sports scenes in the cap of the lighter.

PERIOD 3

#18 – 1951-1959 Painted Models

Zippo produced color-filled illustrations with no background (model #180) engraved drawings from 1951-59. The retail price was $4.00. Motifs included the sail boat, horse, woman bowler (reg.), male bowler (reg.), fisherman, dog with white whiskers, dog (prototype), dog with a hunter in red coat (reg.), golfer running deer, and hunter (prototype).

No. 180 Sport Series

FISHERMAN   BOWLER   SAIL BOAT

HUNTER   DOG   HORSE'S HEAD

BASEBALL PLAYER   GOLFER

#18 A – Sailboat
#18 B – Horse
#18 C – Woman Bowler
    #18 C1 – Woman Bowler (slim)
#18 D – Male Bowler
#18 E – Dog (looking left)
#18 F – Dog (*See page 45, a Sports advertising sheet.*)
#18 G – Prototype Dog (18 G is on the reverse side of 18 D)
#18 H – Baseball Player
#18 I – Hunter
    #18 I1 – 1951 Prototype Hunter
#18 J – Fisherman
#18 K – Golfer
#18 L – 1958 Running Deer (extension of the T&C Sports series). This prototype was done in high polish chrome and was never a production lighter.

If one looks carefully at access #18 L and #27 P one can see why Zippo and its employees (like Wayne Edwards) considered the Town and Country series an extension of the Sports series.

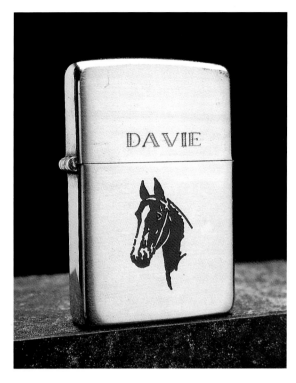

#18 B, $250-$350

#18 A, $250-$350

#18 C, $250-$350

#18 C1, $250-$350

#18 D, $250-$350

#18 E, $250-$350

#18 D, $250-$350

#18 F, obverse, $250-$350

#18 D, $500-$700

#18 F, reverse, $250-$350

48

#18 G, $500-$700

#18 H, obverse, $250-$350

#18 H, reverse

#18 G, $500-$700

#18 I, $250-$350

#18 H, $250-$350

#18 J, $250-$350 without box; $350-$500 with box.

#18 J, $250-$350

#18 L, $700-$1,000 (high polish prototype)

#18 K, $250-$350

#18 II, $700-$1,000 (prototype Hunter)

PERIOD 4

#19 – 1958 Sports Models

These four designs were only produced in 1958 as a set. They include a hunter, golfer, bowler, and fisherman.

#19 A – Hunter
#19 B – Golfer
#19 C – Bowler
#19 D – Fisherman

#19 C, $250-$350

#19 A, $250-$350

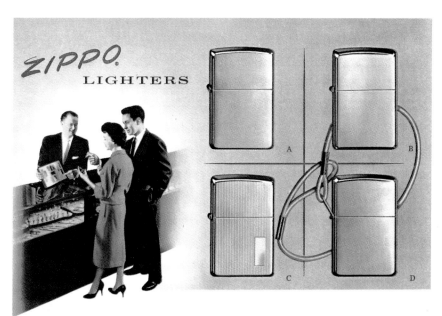

#19 B, $250-$350

PERIOD 5

#20 – 1959-1970

This style had illustrations that were engraved in the lid, as well as having background engraved drawings that were done in color. The skier, trout fisherman, golfer, hunter, male bowler, and snowmobiler were produced during this time period.

Test samples were produced on slims illustrating the downhill skier, fisherman, female bowler, and female golfer. A few were produced for the salesmen to carry in their cases, to see how well they might sell. I highly value the few examples that I have been fortunate enough to acquire.

Factoid: Rudy Bickel (long-time Zippo employee) was the photographer who took photos of Mrs. Harriett Wick (Blaisdell's daughter), as a tinplate for the "slim" female golfer. Few slim examples were ever produced in any of the illustrations.

Factoid: According to Julio Pedine, the oldest member in the art department, in 1960, Mr. Blaisdell asked the art department to try to come up with new Sports illustrations. Julio produced this illustration of a bow hunter stalking a deer. Although this is a wonderful illustration, it was not picked to be part of the series. This is the only example that Julio made. *See access #20 K.*

#20 A – Skier
#20 B – Fisherman (*Note the two illustrations for #20B. Their colors are slightly different.)
#20 C – Golfer
#20 D – Hunter
#20 E – Male Bowler
#20 F – Snowmobiler
#20 G – Male Skier (test sample)
#20 H – Female Bowler (test sample)

#20 I – Female Golfer (test sample)
   #20 I1 – 1980 Female Golfer (replica test sample)
#20 J – 1964 Fisherman (test sample)
#20 K – 1960 Bow Hunter (test sample)

Zippo made some Sports models, during this same time in which there were no illustrations in the lid (#20 L and #20 M). The running deer and deer head are examples. See photos of both deer on Sports advertising seen sheet below.

#20 L – Running Deer (The prototype for this motif has access #21 A)
#20 M – Deer Head
#20 N – 1969 Prototype Hunter
#20 O – 1969 Prototype Hunter
#20 P – 1968 Prototype Hunter (two-sided)

#20 A, $500-$700

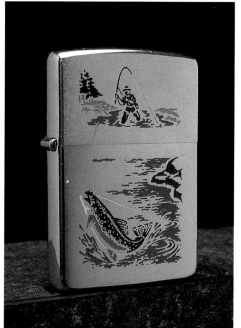

#20 B, $125-$200

$20 B, $125-$200

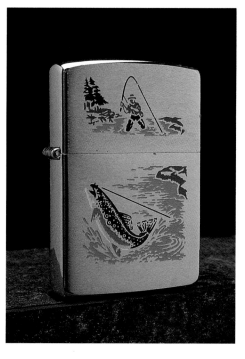

$20 A, $500-$700

#20 B, $125-$200

#20 C, $125-$200

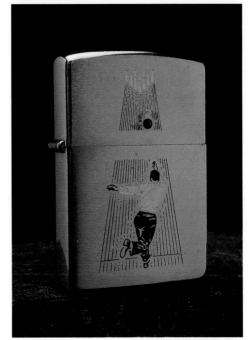

#20 E, $125-$200

#20 D, $125-$200

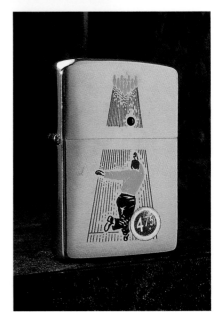

#20 E, $125-$200

#20 D, $125-$200

#20 F, $500-$700

#20 G, $700-$1,000
(test sample)

#20 J, $700-$1,000
(test sample)

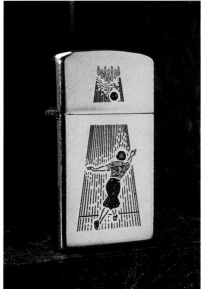
#20 H, $500-$700
(test sample)

#20 I, obverse, $500-$700
(test sample)

#20 H, test sample,
$500-$700

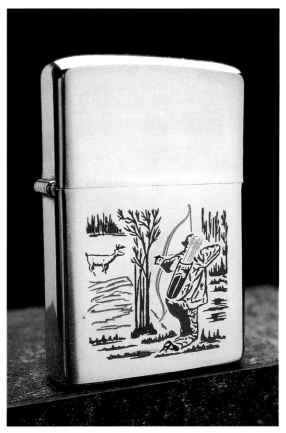
#20 K. There was only one test model of this illustration ever produced. $2500

#20 L, $125-$200

#20 O, $2,000-$2,500
(double sided prototype)

#20 M, $125-$200

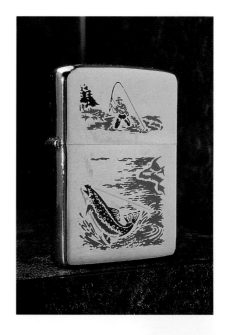

#20 P, obverse,
$125-$200

#20 N, $1,500-
$2,000

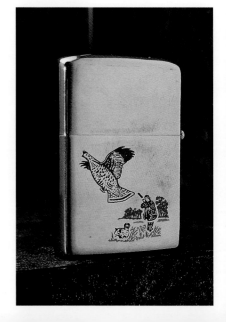

#20 O, reverse

#21 – 1962-1978

Unusual Sports Prototypes & Test Models. These test models are extremely rare and desirable. They are produced like the Moonlander.

#21 A – 1962 Deer. This is a Town and Country transitional prototype illustration. This motif was not silk-screened.

#21 B – 1979 Trout (test sample)
   #21 B1 – 1968 Trout (test sample with advertising logo on the back)

#21 C – 1977 Fisherman (high polish prototype)
   #21 C1 – 1976 Fisherman (brushed finish prototype)

#21 D – 1968 Golfer. This prototype has a high polished finish and is hand painted.
   #21 D1 – 1969 Golfer. This variant prototype has a brushed finish but is not hand painted.
   #21 D2 – 1969 Golfer. This variant prototype has a brushed finish but is not hand painted. Note the differences between #21 D and #21 D2.

#21 E – 1979 Golfer (prototype with blue face)
#21 F – 1978 Prototype Hunter
#21 I – 1968 Trout (test sample with advertising logo on the back)

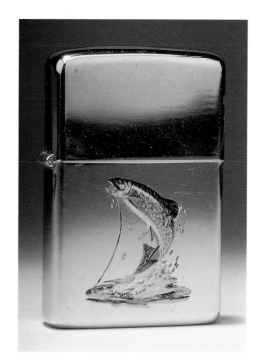

#21 B, $2,500-$3,000 (test sample)

#21 A, $2,500-$3,000 (test sample)

#21 C, $2,500-$3,000 (test sample)

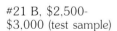

#21 B, $2,500-$3,000 (test sample)

#21 C1, $2,500-$3,000 (test sample)

#21 D, obverse.

#21 E, test sample, $2,500-$3,000

#21 D, reverse, $2,500-$3,000 (test sample)

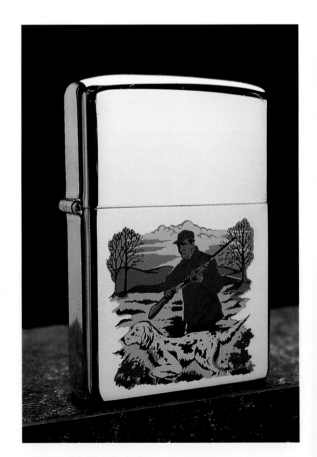

#21 F, $2,500-$3,000 (test sample)

#21 D2, $2,500-$3,000 (test sample)

PERIOD 6

#22 – 1970-81 Models

Zippo designed motifs between 1970 and 1981 that were only found on the bottom half of the case. Zippo produced eight illustrations during this time: fisherman, hunter, snowmobiler, female tennis player, male tennis player, skier, male bowler, and curler (Canada).

Some illustrations had as many as seven colors. The 1974 tennis player is an example.

#22 A – Fisherman. The prototypes for this lighter have access #21 C and #21 C1.
#22 B – Hunter
#22 C – Snowmobiler
#22 D – Female Tennis Player (possible test sample)
#22 E – Male Tennis Player
#22 F – Skier
#22 G – Bowler
#22 H – Hockey Curler – (made for Canada)
#22 I – Golfer *The prototypes for this lighter have access #s 21D, 21D2, and 21E.

#22 E, $75-$125

#22 A, $75-$125

#22 E, $75-$125

#22 C, $75-$125

#22 F, $75-$125

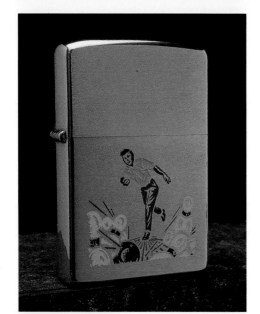

#22 G, $75-$125

PERIOD 7

#23 – 1981-1983 Models

During 1981-83 Zippo produced six models that us[ed] only three or four colors. These models were only produc[ed] from 1981-83. Note that Periods 6 and 8 overlapped w[ith] Period 7. Therefore, other Sports motifs were produc[ed] from during this same time frame. The Snowmobiler w[as] also produced during this period although the picture was modified.

#23 A – Fisherman (with blue sky)
#23 B – Bowler
#23 C – Tennis Player
#23 D – Golfer (This example has 4 white dots to the l[eft] of the golfer)
#23 D1 – 1981 Variant Golfer (This model has 8 wh[ite] dots to the left of the golfer)
#23 E – Hunter (with no leaves [on] trees). The prototype for this ligh[ter] has access number 21 F.)
#23 F – Skier
#23 G – Snowmobiler

#22 I, $75-$125

#23 B, $150-$200

#23 D1, $150-$200

#22 I, $150-$200

#23 E, $150-$200

#23 A, C, D1, E, $150-$200 ea.

PERIOD 8

24 – Circa 1982 to the Present

The motifs in the current Sports series are all illustrated in circular patterns with the exception of the snowmobiler. Examples include the male golfer, male tennis, male tennis, male bowler, male fisherman, male skier, and male hunter.

24 A – Golfer
24 B – Tennis Player
24 C – Bowler
24 D – Fisherman
24 E – Skier
24 F – Hunter
   #24 F1 – 1986 Hunter (prototype)
24 G – Snowmobiler
24 H – 1989 Soccer Player (test model)
24 I – 1986 Slim Woman Bowler
24 J – Bucking Bronco (test model similar to #17 A1)
24 K – Deer Head (test model similar to #20 M)
24 L – Horse Head (test model similar to #18 B)
24 M – Running Deer (test model similar to #20 L)

#24 F1, $700-$1,000

#24 D, $25-$35

#24 I, $40-$60

#24 H, $700-$1,000

$24 J, $700-$1,000

#24 K, $700-$1,000

#24 L, $700-$1,000

#24 M, $700-$1,000

#25 – 1982: GAME SERIES

As I previously stated, the Game Series is really an extension of the Sports series which is the reason why I've placed them here. Examples include: English pointer, English setter, grouse, mallard duck, wood cock, raccoon, bass, bear, deer, flying turkey, pheasant, strutting turkey, running deer, and trout.

Zippo produced two distinct sets of different illustrations of the Game series.

Set 1

#25 A – Running Deer
    #25 A1 – 1986 Running Deer (prototype; never produced in a matte finish and motif is slightly changed)
#25 B – Deer
    #25 B1 – 1987 Deer (prototype; never produced in a grey matte finish)
    #25 B2 – 1986 Deer (prototype; never produced in a black matte finish)
#25 C – Bear
    #25 C1 – 1989 Bear (prototype; in which the bear is brown and not black like the production model)
#25 D – Raccoon
#25 E – English Setter
    #25 E1 – 1989 English Setter (prototype; never produced in a matte finish)
#25 F – English Pointer
#25 G – Flying Pheasant
#25 H – Pheasant
#25 I – Trout
#25 J – Deer (prototype)

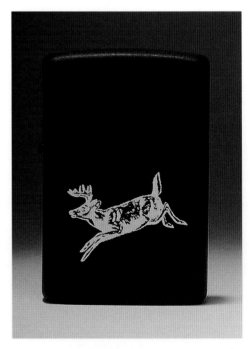

#25 A1, $700-$1,000

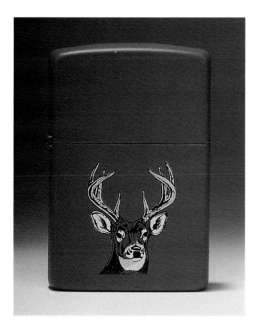

#25 B1, $700-$1,000

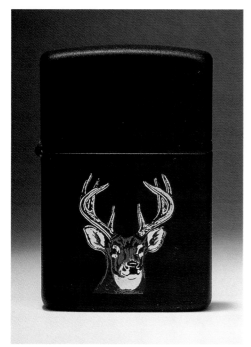

#25 B2, $700-$1,000

#25 E1, $700-$1,000

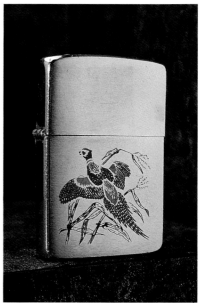

#25 G, $25-$35

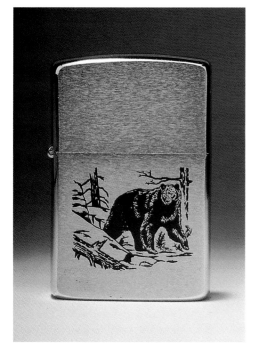

#25 C1, $300-$500

#25 G, $25-$35

Set 2
#26 A – Woodcock
#26 B – Grouse
    #26 B1 – 1989 Grouse (prototype; never produced in a matte finish)
#26 C – 1987 Strutting Turkey (prototype)
#26 D – Flying Turkey
    #26 D1 – 1987 Flying Turkey (prototype; never produced in a matte finish)
#26 E – Strutting Turkey
#26 F – Mallard Duck
#26 G – Bass
    #26 G1 – 1989 Bass (prototype; never produced in a matte finish)

#26 B1, $700-$1,000

#26 G1, $700-$1,000

#25 J, $700-$1,000

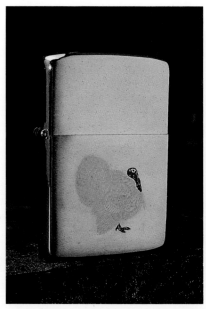

#26 C, $500-$700

#26 D1, $700-$1,000

#26 G1, $700-$1,000

The Town and Country Series

#27 – 1947-1960 TOWN AND COUNTRY Series

The Town and Country series (model #750) began January 1949 as an off-shoot of the Sports Series. Some examples were produced until approximately 1960. The original series consisted of eight lighters: duck, pheasant, geese, horse, trout, setter, lily pond, and sloop. All the illustrations were done by hand. The motifs were cut deep into the metal to help to prevent loss of paint. The illustrations were then air brushed with acrylic enamel using different colors. Next, the final product was electro-baked to increase durability. Zippo called these "baked ceramic enamel" illustrations, at the time. These are considered to be the most beautiful and desirable series to collect. They are very hard to find with 100% paint due to having no metal separating the colors. Because most people carried their lighters in their pockets along with coins, keys, etc. the metal-on-metal contact nearly always chipped the painted surface. As a result, mint quality Town and Country pieces are highly desirable. Today Zippo refers to Town and Country motifs as "paint on paint." Since all Town and Country illustrations are hand painted by different artists, no two illustrations of the same picture are exactly alike.

ZIPPO TOWN AND COUNTRY

The newest, smartest lighter designs in years! A selection of engraved and hand-painted subjects including setter, mallard, geese, trout, sloop, horse, pheasant and lily pond. Each in special, richly-lined gift box.

ZIPPO baked ceramic enamel

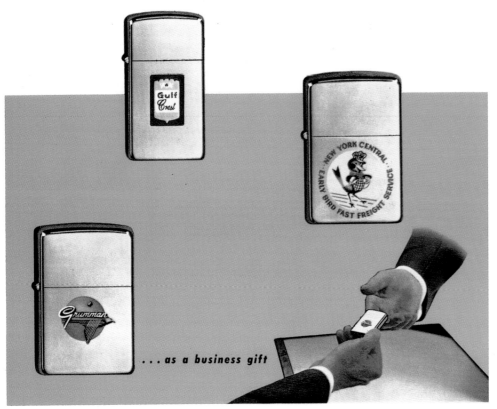

...as a business gift

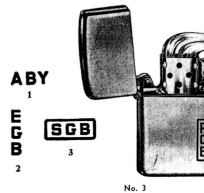

No. 1610 SLIM	QUANTITY	100 to 199	200 to 399	400 and Over
	With One Engraved Surface	$4.10	$4.00	$3.85
	For Each Additional Surface	1.25	1.15	1.00

No. 250 High Polish	QUANTITY	100 to 199	200 to 399	400 and Over
	With One Engraved Surface	$4.10	$4.00	$3.85
	For Each Additional Surface	1.25	1.15	1.00

Town and Country Overview: (Known Benchmarks)

Late 1947-48: Zippo produced many prototypes of Town and Country pictures. Several were sold by salesmen to promote a customer's company.

January, 1949: The Town and Country series was formally put on sale. The trout, sloop, setter, mallard, horse, geese, pheasant, and lily pad were included in the eight original (basic) illustrations @ $7.50 ea. Town and Country motifs were also offered on sterling silver pocket lighters @ $18.50 ea. And on 2nd Model Barcroft table lighters @ $13.50 ea. in all of the eight basic designs. Pocket lighter illustrations were only produced on lighters that had a high polish chrome finish. Some prototype and test model illustrations have appeared on pocket lighters that had a brush finish.

Z112

July, 1949: Town and Country illustrations were produced on the 3rd Model Barcrofts instead of 2nd Model Barcrofts @ $13.50.

September, 1951: Zippo reduced the number of designs from eight to six. Included in the new list were the trout, sloop, setter, mallard, horse, and pheasant @ $7.50 ea. Zippo eliminated both the lily pond scene and the pair of geese from the series, for various reasons.

April, 1952: The price of the Town and Country illustrations was marked up to $8.50. The six basic 1951 designs were still being offered.

November, 1953: Zippo added the sailfish to the previous set of six illustrations. Therefore seven basic pocket lighter designs were offered. Included were the trout, sloop, setter, mallard, horse, pheasant, and sailfish @ $8.50 ea. The sailfish as well as the other motifs could also be purchased on the 3rd Model Barcroft @ $13.50.

Circa 1952: Town and Country designs were offered on 4th Model instead of 3rd Model Barcrofts. It is not known which designs were definitely offered at this time.

1958: There were only three illustrations still being offered that belonged to the original series: trout, mallard, and the pheasant @ $8.50 ea. These three motifs were offered on the 4th model Barcroft as well.

1960: By 1960, Zippo retires the Town and Country series.

Although Zippo officially quit making the Town and Country series in 1960, the technique was still being used as late as 1964. Between 1960 and 1964 Zippo still made a few Town and Country series illustrations. The vast majority of Town and Country illustrations that were produced during this time were test models and prototypes of ships and advertising illustrations which used the traditional "Town and Country" process.

Zippo also produced many transitional pieces which still used the air brush technique. The illustrations of the traditional pieces were "not cut as deep" into the surface of the lighter. These are almost as rare and desirable as the original "deep cut" Town & Countrys. Many of these illustrations are quite beautiful. Also, at this same time, Zippo was also perfecting and implementing its new silk-screen technique. It one looks at a silk-screen illustration carefully using a magnifying glass, one can see a tiny silk-screen "window screen mesh". I call these "silk-screen" lighters. I have contrasted a Town and Country process lighter with a silk-screened surface to show you the difference. Lighters manufactured from 1961-1975 that didn't have metal separating the colors, are highly desirable. One of the reasons is that they are unable to be repainted by Zippo, due to no metal separating the colors. Zippo even made a Town and Country replica of the Lily Pond scene on a slim using the "silk-screen" process. Later, Zippo improved the process, whereas the tiny silk-screen mesh can't be seen even using a 10x magnifier. The "1969 moon lander" is an example of this improved process.

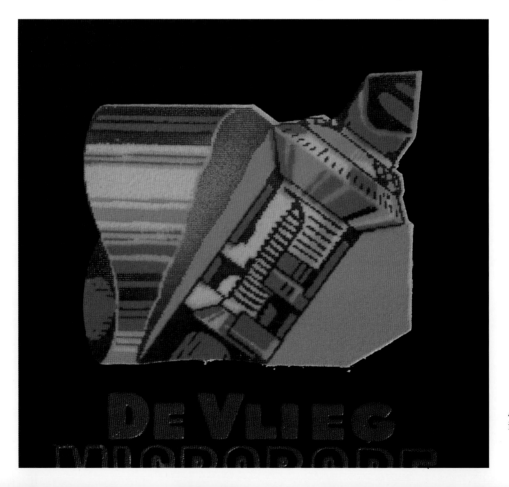

A silk-screened illustration

Airbrushed illustration

#27 A – Horse
#27 B – Setter
#27 C – Sloop
#27 D – Lily Pond
  #27 D1 – 1958 Lily Pond (slim)
#27 E – Pheasant
#27 F – Duck
#27 G – Trout
  #27 G1 – Trout (Prototype variation in which the water is engraved by lines.)
#27 H – Geese
#27 I – 1958 Sailfish
#27 J – 1953 Cow with Wreathe (test model)
#27 K – 1949-50 Bull (test model)
#27 L – 1948-49 Morning Glories (test model)
#27 M – Swans On a Pond (test model)
#27 N – 1949-50 Buffalo (test model) *Extra rare to find with a brush finish.
#27 O – 1949-50 Cherries on a Vine (test model)
#27 P – 1959 Deer Running Away (test model)
#27 Q – Deer Head (test model) (Forerunner of #20M)
#27 R – 1957 Northern Pike (test model)
#27 S – Flying Parrot (test model)
#27 T – 1961 Bull; which says Beewood/Hereford/Farms/Sarasota, Florida; on the reverse side (prototype)

#27 A, $1,200-$1,500

#27 B, $1,200-$1,500

#27 C, $1,200-$1,500

#27 D, $1,400-$1,700

#27 D1, $1,200-$1,500

#27 F, $1,200-$1,500

#27 E, $1,200-$1,500

#27 F, $1,200-$1,500

#27 F, $1,200-$1,500

#27 F, $1,200-$1,500

#27 G, $1,200-$1,500

#27 G1, $1,800-$2,000

#27 G, $1,200-$1,500

#27 H, $1,400-$1,700

#27 G, $1,200-$1,500

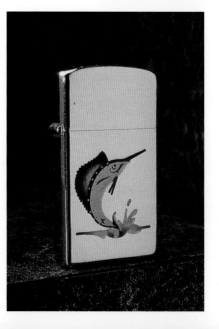

#27 I, $1,400-$1,700

#27 J, $2,500-$3,000

#27 N, $3,000-$3,500

#27 K, $2,500-$3,000

#27 O, $2,500-$3,000

#27 L, $2,500-$3,000

#27 P, $2,500-$3,000

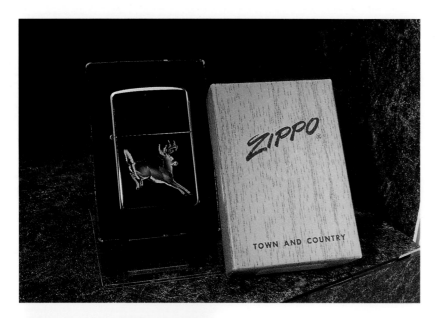

#27 P, $2,500-$3,000. Box is valued at $100-$150.

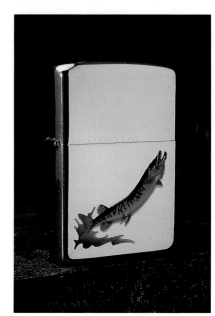

#27 R, $3,000-$3,500

#27 T, obverse, $2,500-$3,000

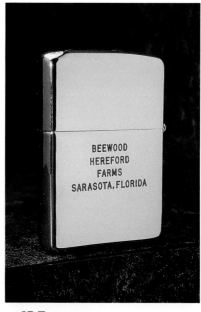

#27 T, reverse

#27 U (slim), $1,200-$1,500

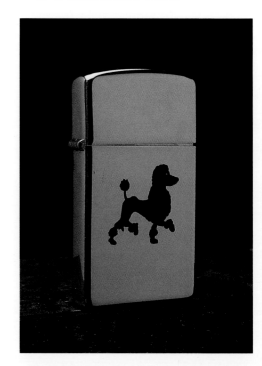

#27 U – 1960 Poodle (slim & regular test models)
#27 V – 1962 Slim Kitten (test model)
#27 W – 1962 Slim Kitten (test model)
#27 X – 1961 Find'em, Fuel'em, Forget'em (test model)
#27 Y – 1957 Bee Hive (prototype)
#27 Z – 1957 Poseidon and Mermaid (test model)

Adv. D.M. Promotion

GIFT OF A LIFETIME by ZIPPO

PERSONALIZED WITH YOUR NAME RANCH AND HERD SIRE...

•

for those buyers you'd like to remember with a LIFETIME GIFT. Not only tops in all 'round lighter performance, but tops, too, as the finest in gift-giving.

•

Especially with your Herd Sire (also identified by name) in life-like ceramics on one side. Your brand design, ranch name and address on the other. (Faithfully reproduced signatures are available on all Zippo models)

•

Zippo is the ONE-ZIP windproof lighter that lights with a ZIP even in wind or rain... is unconditionally guaranteed...

GOLD... a highly red gift for years ears. In 100 quantity 5 *each.

NG SILVER... ive... gift of a e. in 100 quantity *each.

HIGH POLISH... Bright chrome background, dependable finish. In 100 quantity, $5.72 *each.

BRUSH CHROME... Soft-toned, conservative lines always in good taste. In 100 quantity, $5.05 *each.

*Single unit prices slightly higher.

#27 X, $1,200-$1,500

#27 Y, $2,500-$3,000

#27 V, $1,200-$1,500

#27 W, $1,200-$1,500

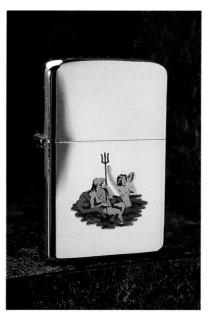

#27 Z, $2,500-$3,000

The War and Beyond: 1943-1960

#28 – 1943-1945 BLACK CRACKLE

The 1943-1945 model was different from the 1942 model in that it had a three barrel hinge. The 1943-1945 insert was also different from the 1942 model in that the steel was wrapped the same direction as other inserts. The insert's flint wheel and chimney were like that of the 1937-1942 models. This model still had a flat bottom that was slightly curved outward. The Zippo logo with the 2032695 patent number was stamped on the insert. The 1942 model has no such logo stamped on the insert. Zippo used this same hinge design on the early model 1946s. There were no insignia models produced in 1943 but Zippo again produced insignias in 1944 and 1945.

#28 A – Plain with black crackle paint
#28 B – Engraved Initials, done by Zippo
#28 C – "In Memory of Ernie Pyle" Zippo (600 produced)
#28 D – Mr. Donald Hyde's "TOKYO" Zippo ("Battlefield Engraved" – only one was ever produced)
#28 E – 1944 and 1945 Insignia Models
#28 F – "Trench Art" Military Emblems
#28 G – Crossed Guns (engraved)

#28 A, $400-$600

Zippo stated that during World War II all models were "Shipped to the Seat of War," but Zippo produced several "special order" models made of steel with a three barrel hinge and a "high polish" finish, that either had a monogram or signature.

#28 H – Above Mentioned Variety – Three barrel steel model with high polished finish, with or without a monogram or signature facsimile.

Sterling "Pocket" models

I have documented four distinct "types" of Sterling Models that were manufactured between 1946 and 1995. I have given these models access numbers 29, 30, 31, and 32. I felt it was better to group them together than to separate them chronologically.

Whether you are trying to date sterling, silver-filled, or gold-filled models take special note of the insert. The insert can be very helpful in dating your lighter, if you know for sure that your lighter has the original insert.

#29 A, $500-$700

#29 – TYPE 1 – 1946 to early 1948 sterling model

Sterling models made during this time span had sterling written under the three barrel hinge in block letters. If the lighter is a 1946 model, the insert is of the "old style." There are seven holes located on each side of the chimney and the flint wheel has horizontal teeth. If the lighter is a late model 1946 or early model 1947 the insert still has seven holes located on each side of the chimney, but the flint wheel has crisscross cut teeth. If the lighter is a late model 1947 or a 1948 model, it has a "new style" insert (like that of the current shape), with eight holes on each side of the chimney. Also, there is a slope of metal connecting the chimney to the flint wheel.

#29 A – Plain
#29 B – Vertical Engine-Turning

#30 – TYPE 2, 1948-1949 sterling

This model had a three barrel hinge. The words "Zippo & Sterling" were written in block print, on the bottom of the lighter. The logo traveled the width (vertically) and not the length (horizontally). The words ZIPPO and MFG are not separated on the insert logo of either model. The insert should have PIM#2 markings, like the regular non-sterling model. The cotton end of the insert is, for the first time, a felt-pad. See figure PP#2. These illustrations are all found on pages 185 and 186.

#30 A – Plain
#30 B – Vertical Engine-Turning

#31 – TYPE 3

Between 1950 to 1954, Sterling Zippo lighters had a five barrel instead of a three barrel hinge. The logo, on the bottom of the lighter, was still stylized in block print and was written vertically. If the insert had a 2032695 pat. Number it could be a late model 1949, 1950, 1951, 1952, or early 1953 model. If the insert had the 2517191 pat. Number with the large patent pending logo, it could be a late model 1953, 1954, 1955, or 1956 model.

#31 A – Plain
#31 B – Vertical Engine-Turning
#31 C – Sterling Prototype (late 1953 or 1954 model)
#31 D – Roy Rogers "Brand"
    #31 D1 – Box that Sterling Roy Rogers came in.

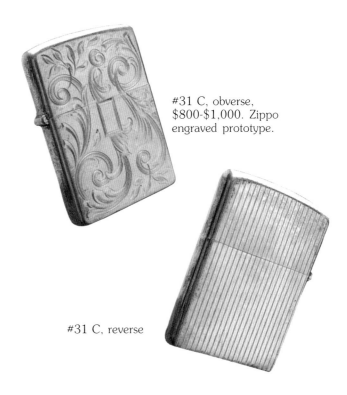

#31 C, obverse, $800-$1,000. Zippo engraved prototype.

#31 C, reverse

#31 D, only one known to exist in any condition. Valued $5,000 in mint condition.

#31 D1, box is valued @ $100-$150.

#33 A1, $350-$500

#33 A1, $350-$500

#33 A1, $350-$500. Box is valued at $150-$200.

#32 – TYPE 4

Circa 1955, Zippo changed the bottom logo on the Sterling model. The logo was produced in script, and written horizontally. This basic model is still produced today.

#32 A – Plain
#32 B – Vertical Engine-Turning
#32 C – Engine-Turned (with beveled rope edges)

Case and Insert Differences: Regular Pocket Models, 1946-1960

Those who feel the need to give every Zippo model an exact manufacturing date are going to be disappointed, due to the fact that Zippo didn't make design changes each year starting January 1, especially in Zippo's early years. Therefore, "model manufacturing years" overlap.

#33 – 1946 Nickel Silver model

Nickel silver material was used to produce Zippo lighters from 1946 to late 1947. Zippo produced three different variations, to be exact.

#33 A – 1946 model:

Case: This model has no chrome plating over the nickel/silver case. The center barrel of the three barrel hinge is longer that the sides. See figure PB#1. The bottom of the case is a little taller than models manufactured today. The length of the bottom of the case is 1' 12/32". Therefore, it is about 1/32 of an inch taller. The markings on the bottom of the case should be PCM#1 or PCM#2.

Insert: This model has the "old style" insert. The flint wheel has horizontal teeth. See figure PF#1. There are seven holes located on each side of the chimney. See figure PC#1. The markings on the insert should be PIM#1. The cotton end of the insert is much like that of a cigarette filter. See figure PP#1. I will call this an Original 1946 or Early 1946.

#33 A1 – Plain
#33 A2 – Generic Advertiser

#33 B – 1946-47 model

This is actually a late model 1946 and early model 1947.

Case: The outer case is chrome plated over nickel silver and the middle hinge barrel is still longer (PB#!). The length of the bottom of the case is still taller (1-12/32"). The markings on the bottom of the case should be PCM#1C or PCM#2C.

Insert: The insert has a new flint wheel with crisscross cut teeth. (Crisscross cut teeth are still used today by Zippo.) See figure PF#2. It still has the "old style" insert. Therefore, there are still seven holes on each side of the chimney (PC#1). The insert must have PIM#! Markings, which are the same as the previous 1946 model. The cotton end of the insert is still much like that of a cigarette filter (PP#1). I will call this a Nickel Silver with Chrome Plating or 1946-47 Model.

#33 B1 – Plain
#33 B2 – Generic Advertiser

#33 C – 1947 model

Case: The outer case was still chrome-plated over nickel silver. This model can be found having two types of hinges. See figures PB#1C and PB#2. The middle hinge barrel of the "early model" PB#1C is a little longer than the hinge barrel of the later PB#2 model. The hinges of both variations were chrome-plated. In 1947 the bottom of the case was shortened 1/32 of an inch. The bottom half of the case for both the 1946 and 1946-47 model was 1-12/32" tall. In contrast, the 1947 model is 1-11/32" tall. See figures PL#1 and PL#2, which compare and contrast the two sizes using front and back photos.

Inserts: This model has the "new style" insert (like that of the current shape). See figure PF#3. Therefore, the chimney had eight holes on each side. See figure PC#2. The cotton end of the insert is still much like that of a cigarette filter, although the 1948-1949 model has a felt pad. I will call this a 1947 Chrome Plated Nickel Silver or 1947 Model. *Zippo used two types of insert logos on the 1947 Model. See figures PIM#1C & PIM#2. Note the location of the word "Zippo" in relationship to the rest of the logo.

#33 C1 – Plain
#33 C2 – Generic Advertiser

#33 D – 1947 Loop model (Loss-proof)

Zippo produced three distinct styles of loss-proof models from 1946-1947 to date. The loop or loss-proof was called the "Tach-A-Loop" according to a June 10, 1949 price list. The Tach-A-Loop was designed and manufactured with the Sportsman in mind. The lanyard prevented his possible keepsake from becoming lost. Loss-proof models are extremely desirable to collect.

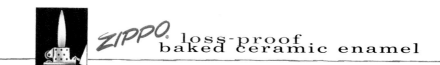

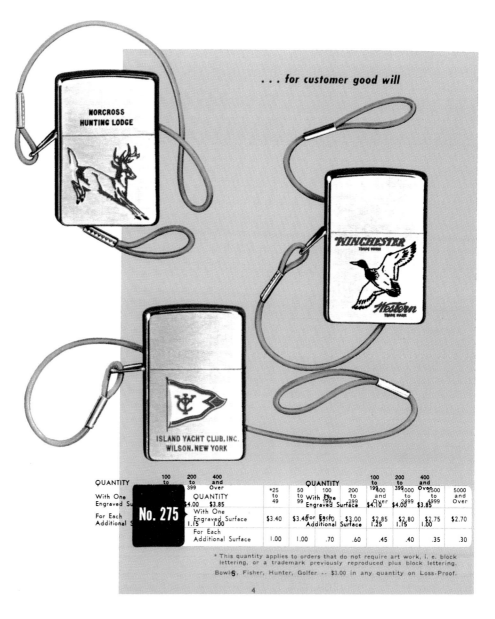

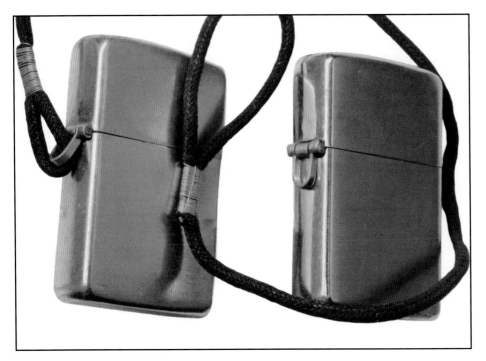

#33 D1, $500-$700 with the lanyard. Extremely rare with lanyard.

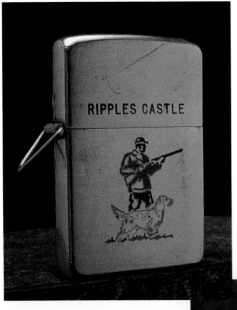

#33 D2, $250-$350. Note the "large" triangular shaped loss proof ring.

#33 D1 – TYPE 1. The ring that attached the lighter's hinge pin to the lanyard was U-shaped from 1946-1947 to about 1951.
#33 D1A – Generic Advertiser
#33 D2 – TYPE 2 – Circa 1952 through 1956 Zippo produced an extra large triangular shaped lossproof ring. The base of the triangle was actually the hinge pin.
#33 D2A – Generic Advertiser
#33 D3 – TYPE 3 – Circa 1957 Zippo changed the size of the triangle and made it smaller. Zippo still produces this same size of triangular ring today.
#33 D3A – Generic Advertiser
#33 D3B – 1959-70 Fisherman

#33 D3, $150-$200 with the lanyard.

#34 – 1948-49 model

Case: This model has a chrome-plated brass case in lieu of a chrome-plated nickel-silver case. This model still has a three barrel hinge. The barrels are of similar length, although the center barrel is still a little longer, as shown above with the 1947 model. This model has the PCM#3 markings on the bottom of the case (three barrel hinge).

Insert: The insert should have PIM#2 markings. The cotton end of the insert is, for the first time, a felt pad. See figure PP#2.

#34 A – Plain
#34 B – Generic Advertiser
#34 C – Red Crackle (test model)
#34 D – Lay's Potato Chips Advertiser
#34 E – BF Goodrich Advertiser
#34 F – Wurlitzer Phonograph Music Advertiser
#34 G – Stainless Steel Case Model. This model is extremely rare and can be identified using a magnet. Stainless steel is actually an alloy of nickel and steel. It will be slightly magnetic but not near as magnetic as a chromed steel case. This model has just enough steel to keep it malleable, so that the case could be formed, possibly in the 410 stainless class. When the lighter opens and shuts it produces its own distinctive sound. If this were a nickel/silver alloy the case wouldn't be magnetic at all.

Note: A magnet is needed to differentiate between access numbers 35-41.

*See pages 184-188 to help identify access numbers 28-38.

#34 D, $175-$250

#34 E, $175-$250

#34 F, $175-$250

#34 C, $1,800-$2,500

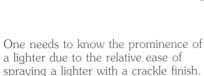

One needs to know the prominence of a lighter due to the relative ease of spraying a lighter with a crackle finish.

#35 B1
This lighter was owned by Wayne Edwards, one of Zippo's 1st employees. The other two names were his two brothers. All three were plant managers at one time or another. I will call this a Zippo/Edwards prototype illustration.
$700-$1,000

#37 C, obverse, $1,000-$1,500. Employee test sample.

#37 C, reverse

#35 – 1949-50 model
This model was different from the 1948-49 model in that it had a five barrel in lieu of a three barrel hinge. It was similar in every other way to the 1948-49 model. It was actually produced from late 1949 through 1950. *See access numbers 40 & 41 concerning the 1950-1957 full leather and 1952-1960 leather wrap models.*
#35 A – Plain
#35 B – Generic Advertiser
#35 B1 – Zippo/Edwards Prototype

#36 – 1951 model
Case: Zippo actually used two types of cases in 1951. Early in 1951, Zippo used the chrome-plated brass case that they had been using since 1948. Mid 1951, with the advent of the Korean War, Zippo switched materials. They began making cases out of chrome-plated steel. This model still had a five barrel hinge like the 1949-50 model. This model also had the 2032695 patent number on the bottom of the "canned" case, although it had three distinct bottom logos. See figures PCM#3 (5 barrel), PCM#4, and PCM#5.

Insert: Zippo used a nickel insert early in 1951 whereas it used a stainless steel insert, mid-1951. Zippo used the PIM#2 markings with the nickel insert whereas they used three distinct markings on the stainless-steel insert. See figures PIM#2S, PIM#3S, and PIM#4S. All insert variations have a felt-pad in the bottom. (PP#2).
#36 A – Plain
#36 B – Generic Advertiser

#37 – 1952-53 model
This model is similar to the preceding model in that it has a chromed-plated steel case, five barrel hinge, and stainless steel insert. It is different from the preceding model in that it can have either a PCM#5-a, PCM#5-b, or PCM#5-c bottom marking. This model must have either the PIM#2S or PIM#3S, insert markings.

The early 1953 model still had the 2032695 patent as you may have found on some of your own lighters, if they happened to be "anniversary" lighters that had a 1953 "anniversary" date Zippo engraved on the case.
#37 A – Plain
#37 B – Generic Advertiser
#37 C – Personalized Zippo Employee Lighter

#38 – 1953 model

This model was the same as the 1953-53 model with the exception of the bottom and insert markings. This model must have the PCM#6, full stamp, bottom marking and the PIM#5 full stamp, insert marking. This model has the PF#3 teeth (on the flint wheel), the PC#2S chimney and holes, and the PB#3 barrels like the 1952-53 model.

#38 A – Plain
#38 B – Generic Advertiser

#39 – 1954-55 model

It is similar to the preceding model in that it has a five barrel hinge, stainless steel insert, and the 2517191 patent numbers with the large patent pending logo. It is different from the preceding model in that the case is made of chromed brass instead of chromed steel. *See access #42 concerning the 1955 gold and silver-filled models.*

#39 A – Plain
#39 B – Generic Advertiser

#38 B, $150-$200

#38 B, $150-$200

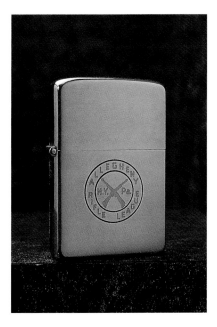

#39 B, $150-$200

#40 – Plain 1950-1957 FULL LEATHER model (#550)

This model has leather covering the entire lighter including the top and the bottom. It was manufactured between 1949 and 1954. It has a gold-leaf border and the word "Zippo" is stylized in gold lettering on the bottom of the lighter. There were available in Genuine Hand-Burnished Calfskin or Imported English Morocco in these colors: red, tan, blue, and green. Again insert markings can be helpful in determining the exact year of production if you are certain that it has the original insert. See "Dating Zippo Lighters" for insert information.

#40 A1, $500-$700

#40 A – Plain (no logo)
    #40 A1 – Red
    #40 A2 – Tan
    #40 A3 – Blue
    #40 A4 – Green
#40 B – Generic Advertiser

#41 – Plain 1952-1960 LEATHER-WRAP

The leather-wrap, also listed as Zippo model number 550, made its debut in 1952. It was manufactured until 1960. The 1952 leather-wrap model was made in conjunction with the 1950 full leather model for a number of years. The leather-wrap model was produced until 1960, whereas the full leather model was produced until 1957. Zippo used a wrap-around application for the lid and bottom. These were available in brown alligator or red, brown, and green reptile. They were also available in black or blue Moroccan leather.

#40 A4, $500-$700

#41 A – Plain (no logo)
    #41 A1 – Brown Alligator
    #41 A2 – Red Reptile
    #41 A3 – Brown Reptile
    #41 A4 – Green Reptile
    #41 A5 – Black Moroccan (Zippo also called this color "coal black.")
    #41 A6 – Blue Moroccan (Zippo also called this color "midnite blue.")
#41 B – Generic Advertiser

Leathercrafted
Luxury styling in rich, genuine leathers. Choice of Morocco, Lizard or Alligator. Black, Blue, Red, Green, Brown
No. 550Z720
Retail $6.00

leathercrafted

In rich Genuine Leathers. Choice of (A.) Brown Alligator, or (B.) Black Morocco.   No. 550 Retail $6.00

#41 A1, $350-$500

#41 A2, $350-$500

#42 – 1955 SILVER & GOLD-FILLED

Silver and gold-filled lighters in the "regular size" were introduced, with the new bottom logo. Of course, Zippo was making gold-filled models in 1937, according to Blaisdell's 1937 *Esquire* advertisement. These two models are the hardest for me to differentiate production years. Zippo never used the yearly code system of dots and slashes on regular size models. Zippo did use the yearly code system on slim models from 1957 until 1961. Circa 1961 through early 1967 Zippo produced slim models with a flat bottom. Circa 1967 through the 1980s Zippo produced slim models that had a canned bottom. The logo on the insert may give you some insight into its exact year of production. One has to be sure that it is the original insert and not merely a replacement to give you a plausible year of production. Zippo used two bottom styles on regular size Zippo lighters. Between 1955 and early 1967 the bottoms of the lighters were flat. Circa 1967 Zippo produced lighters with canned bottoms.

#42 A – Silver-Filled Plain (regular)
#42 B – Silver-Filled Generic Advertising (regular)
#42 C – Gold-Filled Plain (regular)
#42 D – Gold-Filled Generic Advertising (regular)
#42 E – 1967 Gold-Filled Prototype (regular)

#42 E. $500-$700

#43 – 1956 SLIM model

According to Wayne Edwards, Zippo started work on fabricating slim Zippo lighters circa 1955, although Zippo didn't actually start "regular production" of slim lighters until 1956, again with the help of the Backus Novelty Co. of Smethport, PA. Gold and silver-filled slims were also introduced in 1956. The insert had many modifications. The words "pat. Pending" were added to the bottom of the insert. Zippo cut a large, rectangular hole in the back of the chimney. This modification increased the supply of air for easier ignition and better combustion. The cam spring had a "wheel guard" extending from it that kept the flint wheel from getting too hot. It was in the shape of a cobra's head. This piece of metal became brittle over time, and tended to break off easily. Within a short time, Zippo found that the wheel guard impeded the lighters ability to light the wick. Circa 1957, Zippo moved the wick 1/32 of an inch further away from the flint wheel as well as breaking and "grinding off" the wheel guards on the remaining inserts they had in stock. This ensured more effective ignition.

About 1957, Zippo began manufacturing slim inserts without the metal guard. The case, whether it was a 1956 or 1957, had four dots on each side of the "new bottom" logo. Since both 1956s and 1957s have four dots, the easiest way to tell the difference between the two, is to look at the insert to see if it has the wheel guard inside the chimney or not. Most if not all inserts

The "cobra head" wheel guard.

that had the pat. Pending stamp, originally had the wheel guard. One has to look carefully inside the chimney to make sure and see whether their insert originally had the guard or whether it broke off due to mishandling or regular use. Inserts, with the metal guard, are both rare and valuable.

Note: Zippo also made some test sample inserts which had a wheel guard on a regular size insert. These are both rare and valuable to say the least. I personally know of only five that exist.

#43 A – 1956 Plain
#43 B – 1956 Engine turned
#43 C – 1956 Gold-Filled (must have a gold plated insert)
#43 D – 1956 in Original Box
    #43 D1 – Original 1956 Box
#43 E – 1956 Sterling in Original Box
    #43 F – 1956 Generic Advertiser

I listed these unusual slim models here with the 1956 models for convenience.

    #43 G – 1958 Sterling Jockey
    #43 H – 1958 Gold-Filled Jockey
    #43 I – 1960s Sterling Prototype Design
    #43 J – 1958 Sterling Design

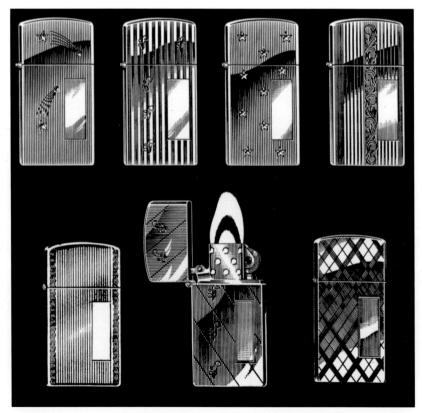

Seven early Slim designs.

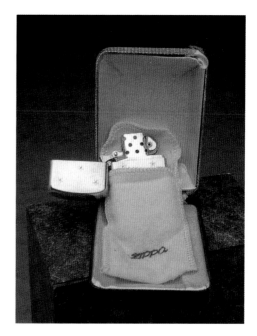

#43 C, $700-$1,000

#43 A, $300-$500

#43 C, open viewed from above. $700-$1,000

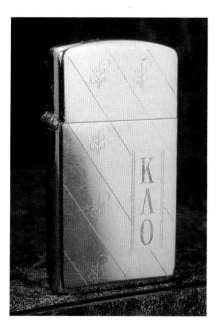

#43 B, $300-$500

#43 D, $500-$700. 1956 in box.

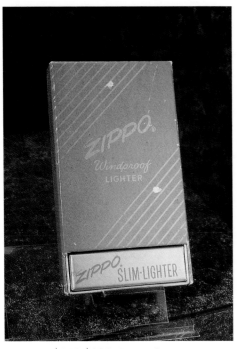

#43 D1, $200-$300. Empty box.

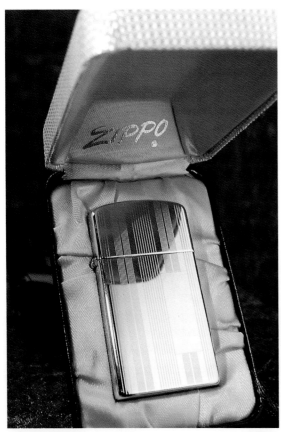

#43 E, $50-$75. Empty box.

#43 G, $175-$250

#43 H, $175-$250

#43 I, $500-$700. Zippo Engraved Prototype.

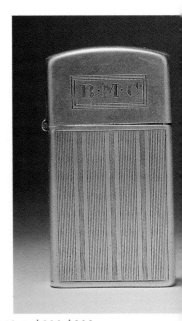

#43 J, $200-$300

#44 – 1958 SLIM 14k GOLD lighter
Zippo started making slim lighters in 14k gold in 1958.

#45 – 1960s VIETNAM Vintage
During the mid 1960s and early 1970s Zippo made some lighters for Vietnam service men that featured "River Division" vessels, helicopters, etc. Both used and mint examples alike are very collectible and desirable. Service men at that time used their Zippo as a canvas to make a statement about themselves or the war effort. Original examples are very desirable.
#45 A – Zippo Engraved" Military Examples
#45 B – "Theater Engraved" Military Examples

Special Interests

#46 – DOG Theme
Zippo began producing a dog theme circa 1960, although dogs were also part of the early Sports series. Some of the original illustrations such as #46 A were hand done by an artisan by the name of Paul Hajdu. Hajdu played an important role in Zippo's success story. Hadju's art work and way of applying paint to the surface of the lighter, were on the cutting edge of technology at that time. Hajdu also used a "special process" in which he applied pictures of actual people's faces to the surface of the lighter. The picture looked like the negative of a photo. Curiously, the pictures cannot be Xeroxed like other illustrations. Hajdu would sometimes do this for friends and Zippo employees. Examples are both rare and extremely desirable.
#46 A – 1966 German Shepherd (prototype)
#46 B – Doberman Pincher
#46 C – Bull Dog
#46 D – Cocker Spaniel (round prototype)
#46 E – 1971 Cocker Spaniel (prototype)
#46 F – 1979 Bloodhound
#46 G – 1992 Golden Retriever
#46 H – 1971 Collie (prototype)
#46 I – 1980 Bearded Collie

#46 A, $300-$500

#46 A, $300-$500

#46 D, $300-$500

#46 E, $300-$500

#46 H, $300-$500

#46 F, $75-$100

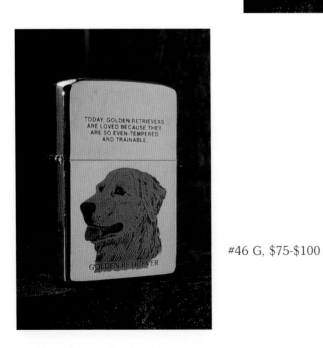

#46 G, $75-$100

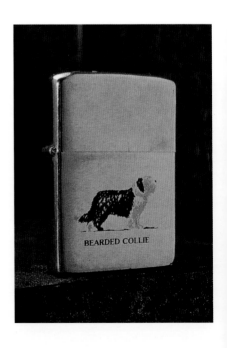

#46 I, $75-$100

#47 – SMILE FACE Set

Zippo produced a Smile motif using five colors of enamel circa 1972. The motifs were produced in green, yellow, orange, brown, and purple enamel in both slims and regulars.

#47 A – Green
    #47 A1 – Green (slim)
#47 B – Yellow
    #47 B1 – Yellow (slim)
#47 C – Orange
    #47 C1 – Orange (slim)
#47 D – Brown
    #47 D1 – Brown (slim)
#47 E – Purple
    #47 E1 – Purple (slim)

In keeping with the 1970s era, Zippo produced motifs of the hang ten design, peace symbol, Snoopy, and even a marijuana leaf design. These illustrations were extremely popular. They were found on tee shirts, sweat shirts, jewelry, lunch boxes, and student note books, as well as the ever popular Zippo lighter. Many of these motifs are again making a come back.

#48 – HANG TEN Motif

The Hang Ten motif was produced in only three colors to my knowledge, red, blue, and purple.

#48 A – Red
    #48 A1 – Red (slim)
#48 B – Blue
    #48 B1 – Blue (slim)
#48 C – Purple
    #48 C1 – Purple (slim)

#47 A, $150-$250

#47 B, $150-$250

#48 C, $150-$250

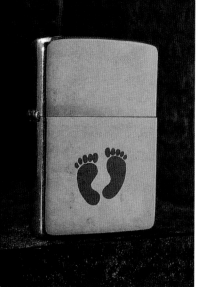

#48 B, $150-$250

#48 C1, $125-$175

#47 E1, $125-$175

#49 – PEACE SIGN Motif
Zippo produced the peace sign on slim lighters in black and white enamel.

#50 – SNOOPY Motif
This motif was primarily produced on slim models.

#51 – MARIJUANA LEAF Motif
This motif was only produced on slim models.

#52 – Vintage CHRISTMAS and VALENTINE Lighters
Holiday lighters have been produced by Zippo almost from its inception. They are rare to find and highly prized. Some examples that I have been fortunate enough to find are: Elf (Rocky), two Valentine lighters, Black Santa, two 1960s Christmas Zippo lighters, 1963 prototype Christmas Zippo, and 1979 prototype Santa. Zippo made a Santa Zippo lighter in 1978. I have seen pictures of the lighter but I have not yet been fortunate enough to both find and purchase one. Approximately 500 were initially produced. In 1993 Zippo manufactured a Black Santa. In 1994, a special order rendition of the 1978 Santa was produced.

#49 A, $150-$250

#51, $200-$300

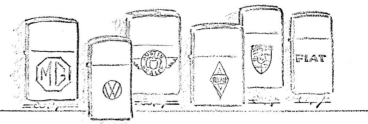

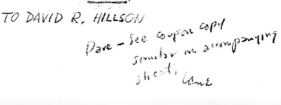

Christmas advertisement layout.

#55 F – 1963 Lockheed Missiles & Space Co. This illustration was produced using the "transitional" Town and Country process, which was not as deep cut but was still air brushed.
#55 G – 1960 Lockheed "Agena Satellite"
#55 H – 1960 ARDC Rocket
#55 I – 1993 Space Odyssey (prototype)

#56 – 1970 (slim) ZODIAC Series

In keeping with the interests of the early '70s, Zippo introduced a Zodiac lighter series. All 12 Zodiac signs were available on slim lighters only. A small number of prototypes were produced on regular size pre-1970 cases. A few regular size examples were also produced during the 1970s and 1980s.

#56 A – Aquarius
#56 B – Pisces
#56 C – Aires
#56 D – Taurus
#56 E – Gemini
#56 F – Cancer
#56 G – Leo
#56 H – Virgo
#56 I – Libra
#56 J – Scorpio
#56 K – Sagittarius
#56 L – Capricorn
#56 M – 1969 Aquarius (prototype)
#56 N – 1969 Aquarius (prototype)
#56 O – 1969 12 Sign (prototype)
#56 P – 1969 12 Sign (prototype)
#56 Q – 1981 Libra (prototype)
#56 R – 1980 Pisces (prototype)
#56 S – 1989 Scorpio (prototype)
#56 T – 1969 Sagittarius (prototype)

#55 H, $100-$150

#55 I, $100-$150

#56 M, $3,000-$5,000

#56 N, $3,000-$5,000

#56 O, $3,000-$5,000

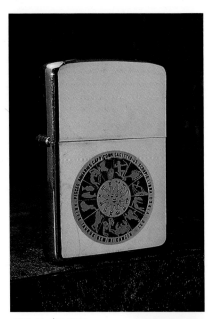

#56 P, $3,000-$5,000

#56 R, $500-$700

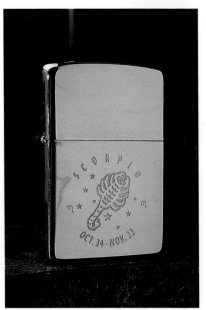

#56 S, $500-$700

#56 T, $1,000-$1,500

#57 – 1970 National Football League Series/Theme

Lighters featuring National Football League helmets and players were introduced as early as 1970. Football motifs can be found on slims as well as regular size models for many of the illustrations.

#57 A – 1972 New England Patriots
#57 B – NFL Regular Size Helmets
#57 C – NFL Slim Size Helmets
    #57 C1 – Bills
#57 D – Pittsburgh Pirates Bucco Power
#57 E – 1976 Super Steelers
#57 F – 1979 Super Steelers
#57 G – 1979 Super Steelers

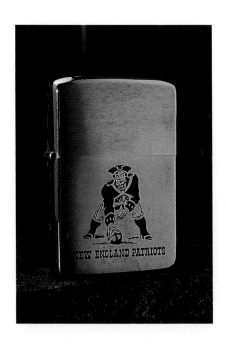

#57 A, $100-$150

Zippo
uts the NFL into a new Hall of Flame!

with a beautiful display of NFL insignia on genuine
o Lighters ... your choice of 26 designs ... each
entic ... each beautifully engraved and color-filled
on choice of brush chrome regular or high polish slim
o lighters ... displayed on colorful plastic shield, it-
personalized with team name... gift boxes for lighters.

y lighter guaranteed to work ... always, or Zippo
fix it free. In 42 years millions of Zippos — but no
— NO ONE has ever paid us a cent to repair a Zippo!

FBL-2250 Display Assortment

3 No. 1610 High Polish Chrome Slim Lighters
  with your choice N. F. L. insignia @ 6.75 ...... $20.25

3 No. 200 Brush Chrome Regular Lighters
  with your choice N. F. L. insignia @ 5.75 ...... $17.25

Total Retail Value $37.50
Dealer Cost per Assortment $22.50
UR PROFIT $15.00

ACTUAL SIZE 10½" W x 10" H

ZIPPO MANUFACTURING COMPANY-BRADFORD, PENNSYLVANIA 16701, IN CANADA: ZIPPO MANUFA... A FALLS, ONTARIO

#57 C1, $50-$75

#57 D, $100-$150

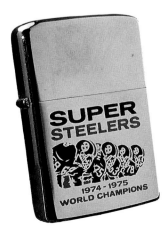

#57 E, $100-$150

#57 F, $100-$150

#57 G, $50-$75

101

#59 A, $150-$200. 1st Zippo key chain (extremely rare)

#59 B, $20-$30. Zippo owns Case Cutlery (extremely hard to find)

#59 B, $20-$30. A Zippo key chain advertising Zippo (extremely hard to find)

#58 – 1972 WOODGRAIN Series
Zippo developed a new and lasting concept in the application of lighter designs when the woodgrain model was introduced. Zippo produced pressure sensitive vinyl appliqués to simulate a woodgrain design. These appliqués were applied on the lighter case. Advertising logos were printed on the woodgrain surface.
#58 A – Plain
#58 B – Generic Advertiser

#59 – 1973 KEY HOLDER
#59 A – Zippo Zip-a-part (1st key ring holder)

#60 – 1974 VENETIAN
#60 A – Plain

#61 – 1976 DENIM Series
In keeping with the clothing fashion at that time, Zippo developed a denim-look lighter. Different motifs were imprinted on the denim-style finish.
#61 A – Plain
#61 B – Yellow Bee (reg.)

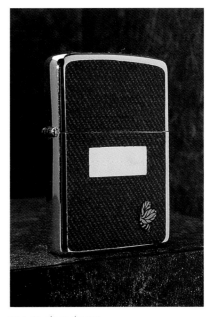

#61 B, $75-$100

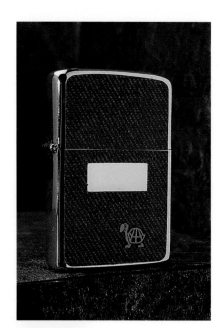

#61 C, $75-$100

#61 C – Green Turtle (reg.)
#61 D – Red Owl (reg.)
#61 E – Rose (slim)
#61 F – Yellow Butterfly (slim)
#61 G – Orange Ladybug (slim)

#62 – 1976 BICENTENNIAL Series

In celebration of the American Bicentennial in 1976, a commemorative Bicentennial lighter was manufactured. Zippo manufactured both a regular and a slim version of the same illustration.
#62 A – Regular Size
#62 B – Slim Size
#62 C – 1976 Prototype

#61 D, $75-$100

#61 G, $40-$60

#61 E, $40-$60

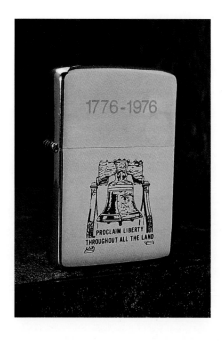

#62 C, $500-$700

#63 – 1977 GOLDEN ELEGANCE Model
#63 A – Plain

#64 – 1977 GOLDEN TORTOISE Model
The handsome Golden Tortoise model was added to the Zippo line in 1977. An acrylic chip was attached to the lighter case transforming it into the attractive tortoise. The Golden Tortoise model is still produced today.
#64 A – Plain

#65 – 1977 SCRIMSHAW
Scrimshaw cases were first made of real scrimshaw and retailed for $37.50. They were test marketed in 1976 and early 1977. Today, scrimshaw cases are made of an acrylic chip and not bone, similar to the ultralite series. To my knowledge there were only three designs that were test marketed in genuine scrimshaw. They were the outrigger (#37), the schooner (#38), and the whaler. The numbers listed are not my access numbers but Zippo's own model numbers. Zippo soon found that "genuine" scrimshaw was too expensive for the average customer therefore, Zippo switched to using acrylic scrimshaw cases, which it still uses today. Zippo used acrylic chips to produce the golden tortoise, ultralite, and scrimshaw series, circa 1977-78.
#65 A – 1977 Outrigger
#65 B – 1977 Schooner
#65 C – Whaler (This motif illustrates a boat full of men harpooning a whale.)

#66 – 1978 ULTRALITE Model
The Ultralite model was added to the general line in 1978. Colorful acrylic chips were used to produce this model.
#66 A – Plain
#66 B – Generic Advertiser
    #66 B1 – 1978 Schultz & Dooley
#66 C – Military Logo
#66 D – 1976 Prototype (This prototype has the new "Flaming I" on the acrylic chip but has the old "Zorro Style Z" on the bottom of the case. Therefore, both logos are on the same lighter.)
#66 E – Pailsey (prototype)

#65 A, $1,000-$1,500

#65 B, obverse, $1,000-$1,500

#66 B1, $50-$75

#65 B, reverse

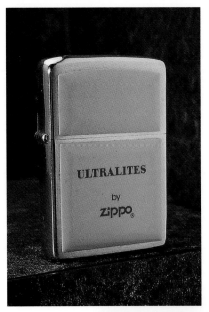

#66 D, $350-$500. 1st Ultralite, a prototype of Zippo Advertising Zippo

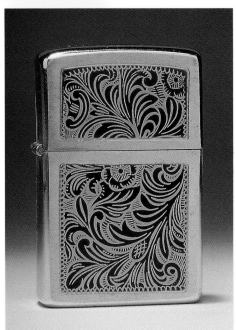

#66 E, $500-$700

#67 – 1979 GOLD ELECTROPLATE
Gold electroplate Zippo lighters were introduced in 1979.
#67 A – Plain
#67 B – Generic Advertiser

#68 – 1979 Political Motifs
In recognition of the 1980 Democratic and Republican National Conventions Zippo produced both a donkey and elephant. (Ronald Reagan was the Republican candidate and Jimmy Carter, the Democratic candidate.) Illustrations were produced on 1979 as well as 1980 cases.
#68 A – 1980 Elephant (production lighters had a silver brushed finish)
    #68 A1 – (high polish silver prototype)
    #68 A2 – (high polish gold plated prototype)
#68 B – 1979 Donkey (production lighters had a silver brushed finish)
    #68 B1 – (high polish silver prototype)
    #68 B2 – (high polish gold plated prototype)

#69 – 1982 50th ANNIVERSARY COMEMORATIVE Lighter
Zippo celebrated its 50th Anniversary with an original design brass lighter, which featured the first diagonal lines

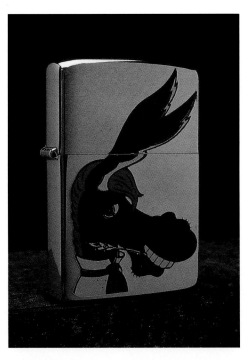

#68 B, $50-$75

cut on the face of the lighter and a seal reading "Fifty Years and Glowing Stronger."
#69 A – 50th Anniversary Commemorative

#70 – 1982 Pipe Lighter
The pipe lighter was different from others in that it had a large hole in contrast to many smaller holes in the chimney. The large hole made it easier to light a pipe by allowing the flame to be drawn through it and into the pipe.
#70 A – Plain
#70 B – 1982 Prototype with Tamper

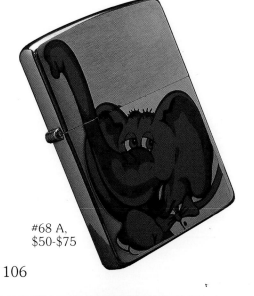

#68 A, $50-$75

#71 - 1982 Patriotic Series/Theme
The Patriotic Series which commemorated the Presidents of the United States officially began in 1982 although Zippo had been making Presidential lighters as early as 1964 beginning with the 1964 John F. Kennedy lighter.
#71 A - Jimmy Carter (regular)
#71 B - Jimmy Carter (slim)
#71 C - 1966 LBJ Signature under Presidential Seal (regular)

#72 - 1983 Solid Brass Lighters
Lighters marked "Solid Brass" on the lid were introduced to the public in 1983. Of course, the 1982 50th Anniversary Commemorative lighter, some Marlboro advertisers, and many prototypes were produced in solid brass (with no finish) before this time, although all of these examples lack the words "Solid Brass" on the lid.
#72 A - Plain
#72 B - Generic Advertiser

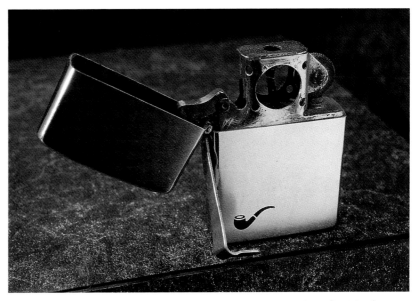

#70 B, $1000-$1,500

#73 - 1984 Powder Coat Model
The black powder coat was introduced which applied electrostatically to the brass case. Soon after four more colors were introduced: blue, burgundy, green, and grey. A year later the imprinted logo and a border were added, Zippo produced all colors in both regular and slim sizes.
#73 A - Blue (with or without border, Zippo logo, or initial panel)
#73 B - Burgundy (with or without border, Zippo logo, or initial panel)
#73 C - Green (with or without border, Zippo logo, or initial panel)
#73 D - Grey (with or without border, Zippo logo, or initial panel)
#73 E - Black (with or without border, Zippo logo, or initial panel)
#73 F - Generic Advertiser

#70 B, $1,000-$1,500

#74 - 1987 (Circa 1987-91) PRESIDENTIAL Set
Two lighters made up the set.
#74 A - 1980s Bush-Gorbachev
#74 B - 1980s Reagan-Gorbachev (A "Special Order" replica of this lighter was produced in the 1990s.)

#75 - 1987 ELVIS Series
Zippo commemorated the "King's" birthday with four different slims & three different reg. Size Zippos. Access numbers #75 H, #75 H1, #75 I, & #75 J are test models.
#75 A - Regular Size (colored face)

#75 H, $750-$1,000

#75 H1, $750-$1,000

#75 B – Regular Size (circular pattern, high polish chrome)
#75 C – Regular Size (circular pattern with black paint)
#75 D – Slim (colored face)
#75 E – Slim (circular pattern, high polish chrome)
#75 F – Slim (circular pattern with black paint)
#75 G – Slim (standing Elvis with black paint)
#75 H – 1982 Bust (test sample)
    #75 H1 – 1982 Gold-Plated Bust
#75 I – 1990 Test Sample
#75 J – 1978 Standing Elvis (test sample)

*Note: Zippo added the 1932 replica series to it's collection in 1988. The bottom of the lighter says "Original 1932 Replica." This lighter, in its original gray box, trades in the $225-$300 price range.

#76 – 1990 Anheuser-Busch Series
Done in both high polish brass and high polish chrome.

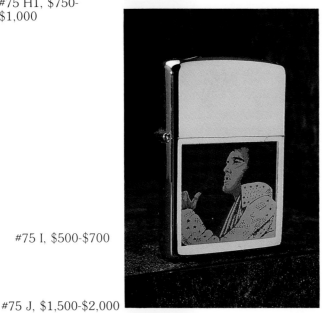

#75 I, $500-$700

#75 J, $1,500-$2,000

#76 A – Michelob

#76 E – Natural Light

#76 B – Bud Light

#76 F – O'Doul's

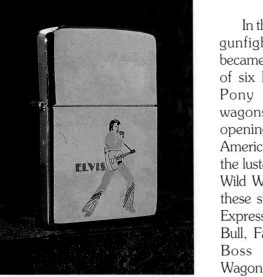

#77 – 1990 WILD WEST Series

In the Wild West Indians, gunfighters, and sheriffs became legends. This series of six lighters depicts the Pony Express, covered wagons, and the railroad opening new territories in America. All examples had the luster etched finish. The Wild West series contained these six illustrations: Pony Express 1880, Chief Sitting Bull, Famous Sheriff, Trail Boss 1890, Conestoga Wagon 1840, and Western Attire.

#76 C – Busch

#76 G – Budweiser

#76 D – Bud Dry

#76 H – A-B Eagle

#77 A – Pony Express
#77 B – Chief Sitting Bull
    #77 B1 – 1995 Chief Sitting Bull (prototype)
#77 C – Famous Sheriff
#77 D – Trail Boss
#77 E – Conestoga Wagon
#77 F – Western Attire

#78 – 1990 PRESIDENTIAL Series

Zippo honored six popular presidents with the President Series. Included in the Series were Ronald Reagan, John Kennedy, Dwight D. Eisenhower, Franklin D. Roosevelt, Abraham Lincoln, and George Washington.
#78 A – Ronald Reagan
#78 B – John Kennedy
    #78 B1 – John F. Kennedy (This prototype illustration was never produced on the 1932 replica for the series.)
#78 C – Dwight Eisenhower
#78 D – Franklin D. Roosevelt
#78 E – Abraham Lincoln
#78 F – George Washington

#77 B1, $300-$500

#79 – 1990 FABULOUS 1950s Series

Four lighters make up the set: 1954 Corvette, 1955 Cadillac, 1956 Thunderbird, and 1957 BelAir.
#79 A – 1954 Corvette
#79 B – 1955 Cadillac
#79 C – 1956 Thunderbird
#79 D – 1957 BelAir

#80 – 1990 SUPER 1960s Series

Four lighters make up the set: 1964 GTO, 1965 Corvette, 1965 Mustang, and 1967 Camaro.
#80 A – 1964 GTO
    #80 A1 – 1964 GTO (prototype on a 1990 lighter)
    #80 A2 – 1964 GTO (prototype done in a matte finish on a 1990 lighter)
#80 B – 1965 Corvette
#80 C – 1965 Mustang
    #80 C1 – 1965 Mustang (prototype on a 1989 lighter) Regular production issue was never produced in blue matte.
#80 D – 1967 Camaro
    #80 D1 – 1967 Camaro (prototype)

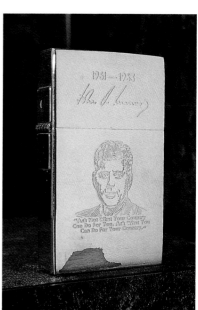
#78 B1, $400-$500

#81 – 1991 CIVIL WAR Series

The Civil War Collection featured beautiful color illustrations of those courageous soldiers who fought for the Union and Confederate armies of the American Civil War. The Union Army had four representative motifs which included the Militiaman, Infantryman, Cavalry Trooper, and Petty Officer. The Confederate Army also

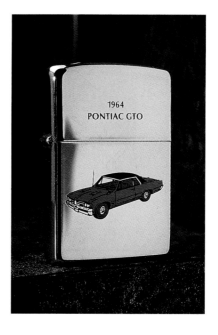
#80 A1, $300-$500

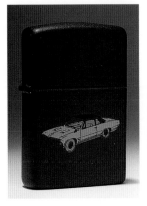
#80 A2, $300-$500

#80 C1, $300-$500

#80 D1, $200-$400

#83 C, $1,500-$2,000

#83 D, $1,500-$2,000

#83 E, $1,500-$2,000

had four representative illustrations which included the Cherokee Rifleman, Calvary Trooper, Infantryman, and Prison Guard.
#81 A – Militiaman (Union Army)
#81 B – Infantryman (Union Army)
#81 C – Cavalry Trooper (Union Army)
#81 D – Petty Officer (Union Army)
#81 E – Cherokee Rifleman (Confederate Army)
#81 F – Cavalry Trooper (Confederate Army)
#81 G – Infantryman (Confederate Army)
#81 H – Prison Guard (Confederate Army)

#82 – 1991 AMERICAN CLASSICS Series

Four lighters make up the American Classics series; USA on Map/Flag, Old Glory, Eagle/USA, and Stars & Stripes.
#82 A – USA on Map/Flag
#82 B – Old Glory
#82 C – Eagle/USA
#82 D – Stars & Stripes

#83 – 1992 60th Anniversary Lighter

In 1992, during its 60th Anniversary, Zippo introduced the 60th Anniversary Commemorative Lighter in a collectible tin. Customer and collector response was so great that Zippo has continued with a new collectible lighter each year. The 1992 Limited Edition (356, 179 lighters), 60th Anniversary Lighter (ZOR #MC60) is the first in the annual "Lighter of the Year" series of collectibles. The 60th Anniversary Commemorative Lighter was midnight chrome with a pewter 60th Anniversary emblem. Zippo also manufactured approximately 1,200 60th Anniversary "Sterling with Gold Inlay" Zippo lighters and gave them to their employees as a gift of appreciation.
#83 A – 60th Anniversary Commemorative (production lighter)
#83 B – 60th Anniversary "Sterling w/Gold Inlay" (production lighter)
#83 C – 60th Anniversary Sterling Prototype
#83 D – 60th Anniversary Sterling Prototype w/GGB Initials
#83 E – 60th Anniversary Commemorative Edition Sterling Prototype

#84 – 1992 60th Zippo Anniversary Series

Zippo celebrated its 60th Anniversary with a set of six lighters. Included in the set were replicas illustrating: the 5th, 10th, 25th, 40th, 50th, and 60th anniversaries.
#84 A – 5th Anniversary
#84 B – 10th Anniversary
#84 C – 25th Anniversary
#84 D – 40th Anniversary
#84 E – 50th Anniversary
#84 F – 60th Anniversary

*Original Anniversary Zippo Lighters begin with the 20th Anniversary Model. Instead of separating and covering the original anniversary models throughout the text, I grouped them all together here. Access #s 84G-84I are original anniversary models. The original 60th Anniversary model has access #83 A.
#84 G – 25th Anniversary
#84 H – 40th Anniversary
#84 I – 50th Anniversary

CIVIL WAR
Zippo Souvenir Lighter Display

This latest series of Zippo lighters brings back the haunted memory of the Civil War. The collection features beautiful illustrations of those courageous soldiers who fought for the Union and Confederate armies

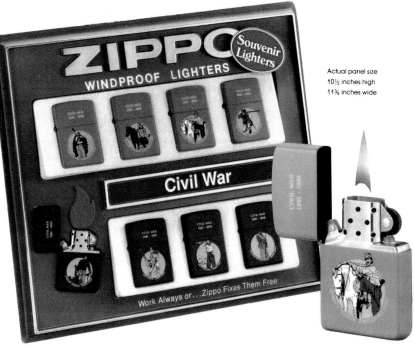

Actual panel size
10½ inches high
11¾ inches wide

Union Army (blue matte)
- Militiaman #220CW-600
- Infantryman #220CW-601
- Cavalry Trooper #220CW-602
- Petty Officer #220CW-603

Confederate Army
(grey matte)
- Cherokee Rifleman #223CW-604
- Cavalry Trooper #223CW-605
- Infantryman #223CW-606
- Prison Guard #223CW-607

		Suggested Retail
CW-8	**Civil War Assortment Panel** Includes eight (8) matte finish Zippo lighters with Confederate and Union civil war soldiers art	$127.95
CWB-8	**Civil War Assortment Panel** Includes two (2) each blue matte finish Zippo lighters with Union Army soldiers art.	127.95
CWG-8	**Civil War Assortment Panel** Includes two (2) each grey matte finish Zippo lighters with Confederate Army soldiers art	127.95
	Individual Lighters Featuring Civil War Soldiers Please specify designs from assortment mentioned above (with gift box).	15.95 each

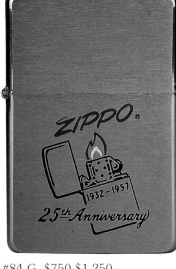

#84 G, $750-$1,250

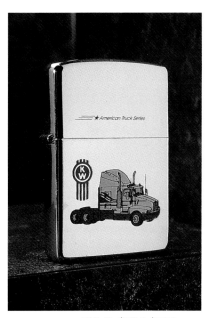

#85 E, $500-$750

#85 – 1992 PETERBILT TRUCK Series

Zippo saluted the eighteen wheeler in 1992 with the Peterbilt Truck series. Four lighters comprised the series: Freightliner, Kenworth, White GMC, and Peterbilt. (I have four "practice" motifs that were produced in 1991, when Zippo's Art Department was designing the series.)

#85 A – Freightliner
#85 B – Kenworth
#85 C – White GMC
#85 D – Peterbilt
#85 E – 1991 Kenworth (prototype)
#85 F – 1991 International (test sample)
#85 G – 1991 Peterbilt (prototype)
#85 H – 1991 Mack (test sample)

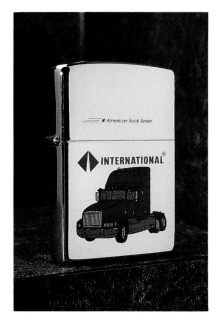

#85 F, $500-$750

111

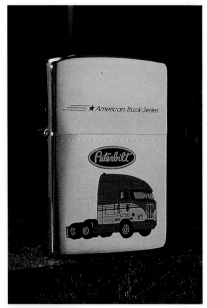

#85 G, $500-$750

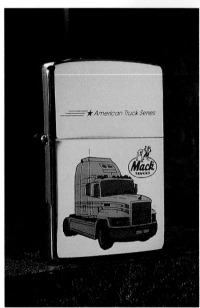

#85 H, $500-$750

#89 B1, $500-$750

#86 – 1992 BORN TO RIDE Series
The fascination...the power...the excitement of motorcycle riding was highlighted on eight lighters in the Born to Ride series. Zippo produced eight cycle designs in both high polish and black matte. Included in the series were: Do It, The Warrior, Soaring Thunder, Café Racer, The Reaper, Maiden America, Rebel Daze, and Born to Ride (This series was still being produced in 1994 in both high polish chrome and black matte.)
#86 A – Do It
#86 B – The Warrior
#86 C – Soaring Thunder
#86 D – Café Racer
#86 E – The Reaper
#86 F – Maiden America
#86 G – Rebel Daze
#86 H – Born to Ride

#87 – 1993 GEOMETRICS COLLECTION
Cool and contemporary, the Geometric Collection featured an exotic group of eight Zippo lighters. Included were: Mobile Ceramic, Blue Mesh, Orange Grid, Assorted symbols, Symmetric Circles, Varied Geometrics, Cartoon Abstract, and Modern Abstract.
#87 A – Mobile Ceramic
#87 B – Blue Mesh
#87 C – Orange Grid
#87 D – Assorted symbols
#87 E – Symmetric Circles
#87 F – Varied Geometrics
#87 G – Cartoon Abstract
#87 H – Modern Abstract

#88 – 1993 VARGA GIRL
The Varga Girl, Windy, was introduced in 1937 in Zippo's first national ad. The 1993 collectible "Lighter of the Year" featured Windy cast in pewter and attached to the lighter as a two surface emblem. The 1993 Limited Edition (459,924 lighters), Varga Girl (ZOR #250YG), is the second in the annual "Lighter of the Year" series. Windy was also pictured on the lid of the collectible tin.
#88 A – Varga Girl

#89 – 1993 VINTAGE AIRCRAFT
World War II produced some of the world's finest aircraft from the American P-51 Mustang to the Japanese Zero. To commemorate the famous planes of that era Zippo produced eight illustrations. This series includes A6M Zero, JU87 Stuka, F6F Hellcat, F4U Corsair, P47 Thunderbolt, B24 Liberator, P51 Mustang, and Curtiss P-40.
#89 A – A6M Zero
#89 B – JU87 Stuka
    #89 B1 – 1992 JU87 Stuka (prototype)
#89 C – F6F Hellcat

#89 D – F4U Corsair
    #89 D1 – 1992 F4U Corsair (prototype)
#89 E – P47 Thunderbolt
#89 F – B24 Liberator
#89 G – P51 Mustang
#89 H – Curtiss P-40

#90 – 1993 CORVETTE Series

The Corvette Collection commemorated the 40th Anniversary of the popular American sports car. Eight lighters made up the Series: 1953, 1957, 1963, 1969, 1978, 1986, and 1991 models.

#90 A – 1953 Corvette
#90 B – 1957 Corvette
#90 C – 1963 Corvette
#90 D – 1969 Corvette
#90 E – 1978 Corvette
#90 F – 1986 Corvette
#90 G – 1991 Corvette
#90 H – 40th Anniversary Corvette

#90 A, $25-$30

#89 D, $500-$750

BARRETT-SMYTHE COLLECTION: 1993-1994

The designers of Barrett-Smythe, Ltd. joined Zippo Manufacturing in producing collectible art using the Zippo windproof lighter as a canvas. An initial selection of 38 Barrett-Smythe designs was introduced in 1993. The collection illustrated endangered animals, animal friends, backyard insects, and L'Art Classique.

Four dinosaurs of a long-ago world were captured in three-dimensional full-face emblems. The giant reptiles were reproduced in pewter on a high polish chrome lighter and in antique brass on a solid brass lighter.

#91 – 1993-94 DINOSAURS

Four giant reptiles were reproduced in pewter on a high polish chrome lighter, and in antique brass on a solid brass lighter. Included were: T. Rex, Triceratops, Dimetrodon, and Stegosaurus.

#91 A – T. Rex
#91 B – Triceratops
#91 C – Dimetrodon
#91 D – Stegosaurus

#92 – 1993-94 ANIMAL FRIENDS

Wild or domestic, animals enrich our lives in many ways. Zippo produced six front-and-back designs of wild and domestic animals on midnight chrome lighters. These include the zebra, black leopard, tiger, pig, kangaroo, and ostrich.

#92 A – Zebra
#92 B – Black Leopard
#92 C – Tiger
#92 D – Pig
#92 E – Kangaroo
#92 F – Ostrich

#93 – 1993-94 BACKYARD INSECTS

Barrett-Smythe produced six designs depicting insects on a midnight chrome finish. They include the praying mantis, bee, lightning bug, spider, butterfly, and ladybug.

#93 A – Praying Mantis
#93 B – Bee
#93 C – Lightning Bug
#93 D – Spider
#93 E – Butterfly
#93 F – Ladybug

#94 – 1993-94 ENDANGERED ANIMALS Series

Zippo's concern for the future of the earth and all its species prompted them to commemorate six endangered animals. These endangered animals make up this set: manatee, whale, elephant, giant panda, gorilla, and rhinoceros. All were done in midnight chrome.

#94 A – Manatee
#94 B – Whale
#94 C – Elephant
#94 D – Giant Panda
#94 E – Gorilla
#94 F – Rhinoceros

Slim Collection: The Slim Collection contains eight lighters which were all done in midnight chrome: butterfly, ladybug, ladybug shadow, roach, cow, spider, cat, and horse.

#94 G – Butterfly
#94 H – Ladybug
#94 I – Ladybug Shadown
#94 J – Roach
#94 K – Cow
#94 L – Spider
#94 M – Cat
#94 N – Horse

#95 – 1993-94 L'ART CLASSIQUE

This series depicts four distinctive and beautiful periods of art. Zippo's midnight chrome lighter provides the background on which to display Barrett-Smythe's interpretations of these styles; Victorian, Modern Art, Art Deco, and Art Nouveau.

#95 A – Victorian
#95 B – Modern Art
#95 C – Art Deco
#95 D – Art Nouveau

#96 – 1994 WILDLIFE

This series depicts seven animal scenes: attacking eagle, eagle sheltering, eagle poised for flight, deer, tiger, lion, and leopard.

#96 A – Attacking eagle
#96 B – Eagle Sheltering
#96 C – Eagle Poised For Flight
#96 D – Deer
#96 E – Tiger
#96 F – Lion
#96 G – Leopard

#97 – 1994 TOLEDO Collection

The Toledo Collection featured the ancient art of damascene engraving. Damascene is the technique of encrusting gold, silver, or copper wire on a surface of iron, steel, or bronze. This technique reached its pinnacle in design and beauty through the craftsmanship of the artisans of Toledo, Spain, Toledo steel, the base for these engravings, is scored with fine lines carved just deep enough into the surface to accept the thin gold wire which is inlaid by hand. The gold is affixed by gentle hammering and then the steel is oxidized to an intense black. The design can then be worked into bas-relief and finished by polishing. For this collection Zippo has commissioned four damascene designs fabricated in Toledo, Spain from 18k gold and steel. Two of the designs feature intricate geometric shapes: the Octastar and the 12-point star. The third depicts a matador in the bull ring and the fourth portrays Don Quixote and his faithful companion Sancho Panza. Each high-polish brass lighter is decorated on the front and back sides with damascene plaques. Toledo lighters capture the beauty of this ancient art.

#97 A – Toledo Octastar (8 pt. Star)
#97 B – Toledo – Matador
#97 C – Toledo – D Quixote
#97 D – Toledo (12 pt. Star)
#97 E – 1992 Slim Test Model
#97 F – 1992 Slim Test Model

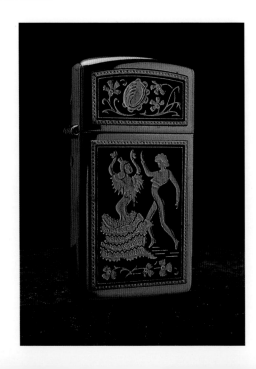

#97 E, $400-$600. Prototype.

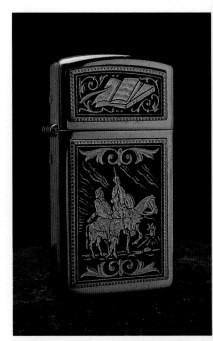

#97 F, $400-$600. Prototype.

#98 – 1994 **Anheuser-Busch Series**

This series is produced in both high polish brass and high polish chromed.
#98 A – Michelob
#98 B – Bud Dry
#98 C – Natural Light
#98 D – Busch
#98 E – Budweiser
#98 F – O'Doul's
#98 G – Bud Light
#98 H – A-B Eagle

#99 – 1994 **SOUVENIR TRUCK Series**

Zippo saluted the king of the road, the eighteen wheeler, in the Souvenir Truck Series. The power of the big rigs was illustrated in seven different full-color designs on black matte lighters. Included in the series were: Knight Rebel, Road Winder, Black Classic, American Spirit, Call of the Wild, Freedom Road, and Born Bad.
#99 A – Knight Rebel
#99 B – Road Winder
#99 C – Black Classic
#99 D – American Spirit
#99 E – Call of the Wild
#99 F – Freedom Road
#99 G – Born Bad

#99 G1, $200-$300

#99 G1 – Born Bad (prototype never produced in high polish chrome.)

#100 – 1994 **STARGATE Series (3)**

Stargate captures the world's imagination with an intriguing adventure involving travel through space to unlock the secrets of ancient Egypt. Zippo joined the adventure by duplicating three mysterious Stargate graphic symbols; the Anubis, Eye of Ra, and Ra Mask, on black matte windproof lighters.
#100 A – Anubis
#100 B – Eye of Ra
#100 C – Ra Mask

Zippo entered the world of motorsports in 1993 with the sponsorship of the U.S. Vintage Grand Prix of Watkins Glen. New in 1993 was the Zippo Motorsports Collection featuring popular NASCAR drivers and Winston Cup graphics. Also during 1993 Zippo commemorated the breaking of an 83-year old tradition at the Indianapolis Motor Speedway as NASCAR drivers took the track in the Inaugural Run of the Brickyard 400. In 1994, Zippo signed on with Team Kendall as an associate sponsor of Sabco Racing's #40 NASCAR entry. In tribute to Watkins Glen and the U.S. Vintage Grand Prix, Zippo created a vintage collection of 13 special edition products imprinted with one of three Watkins Glen logos that could be custom ordered. None of the thirteen special edition products were stock items.

#101 – 1993-1994 MOTORSPORTS COLLECTION

The Motorsports Collection consists of two racing sets that were produced in 1993-94: the Jeff Gordon series and the Winston Cup/NASCAR and Watkins Glen set. I grouped the two together.

JEFF GORDON Series (2 lighters): This series includes both the Brickyard and Brickyard 400 in tins.
#101 A – Brickyard
#101 B – Brickyard 400

WINSTON CUP Series: This Motorsports collection contained eight lighters.
#101 C – Jeff Gordon
#101 D – Dale Jarrett
#101 E – Kenny Wallace
#101 F – Morgan Shepard
#101 G – Mark Martin
#101 H – Kyle Petty
#101 I – Winston Cup
#101 J – Winston Cup (Eagles)

#102 –1994 SCULPTURED Series

This series includes: cat, elephant, bass, horse head, race car, ram head, and saddle.
#102 A – Cat
#102 B – Elephant
#102 C – Bass
#102 D – Horse Head
#102 E – Race Car
#102 F – Ram Head
#102 G – Saddle

103 – 1994 SCRIMSHAW Series

Zippo's Scrimshaw Collection captures the beauty of a traditional art form. Every lighter in the collection features an ivory-toned panel etched with simulated scrimshaw art. Made of acrylic, each panel is decorated with a drawing reminiscent of yesteryear. In 1994 Zippo produced these six designs: whale, walrus, ship, pipe lighter ship, Statue of Liberty, and liberty eagle.

#103 A – Whale
#103 B – Walrus
#103 C – Ship
#103 D – Ship Pipe Lighter
#103 E – Statue of Liberty
#103 F – Liberty Eagle

#104 – 1994 BLACK BEAR scrimshaw

Case knife with lighter, limited edition of 1000.
#104 A – Black Bear

#105 – 1994 BUFFALO BILLS Anniversary Lighter
#105 A – Buffalo Bills Anniversary Lighter (regular)

#106 – 1994 CHRISTMAS Lighters
Zippo created four designs depicting traditional Christmas motifs. They include an Elf, Santa Claus, Santa with Reindeer, and Christmas Tree. A limited edition of 25,000 lighters, of each style, were produced by Zippo.

#106 A – Elf
#106 B – Santa Claus
#106 C – Santa with Reindeer
   #106 C1 – 1994 Santa with Reindeer (This is a prototype. The final product was never made with a brass finish.)
#106 D – Christmas Tree

#106 A, $30-$40

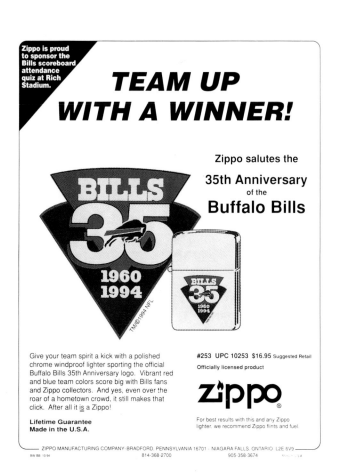

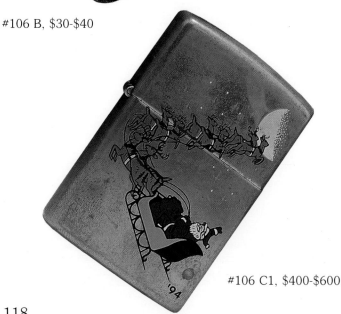
#106 B, $30-$40

#106 C1, $400-$600

118

#107 – 1994 WOODSTOCK Series

Two lighters comprise this series: Then (One Dove) which commemorated the first Woodstock and Now (Two Doves) which commemorated the second Woodstock.

#107 A – Then (Says "Woodstock 3 days of Peace and Music")

#107 B – Now (Says "2 More Days of Peace and Music")

#107 C – Now (variant has red and green colors)

#108 – 1994 D-DAY Lighter

Zippo saluted the 50th Anniversary of D-Day with its 1994 Commemorative lighter. Its black crackly surface replicated the World War II finish, and the antique brass emblem was patterned after the sleeve patch worn by Allied troops. General Dwight Eisenhower's inspirational June 6, 1944 message to the troops is reprinted inside the lid of the round tin in which it comes. With the D-Day commemorative lighter, Zippo honors the 50th Anniversary of the greatest military operation the world has ever known. It replicates the original World War II black crackle finish and is emblazoned with a special D-Day insignia. The 1994 Limited Edition (334,757 lighters), D-Day Lighter (ZOR #236DD) is the third in the annual "Lighter of the Year" series of collectibles.

#108 A – D-Day Lighter

#109 – 1994 ALLIED HEROES Set

Zippo introduced the Allied Heroes Set in 1994 which saluted legendary war heroes. The Allied Heroes Collection depicted four high-polish brass lighters with engraved portraits of World War II military leaders: General Charles de Gaulle, General Bernard L. Montgomery, Lt. General Omar N. Bradley, and General Dwight D. Eisenhower. Zippo also included a complimentary solid brass key ring etched with the special D-Day emblem. The set came in a round metal tin decorated with graphics depicting the beaches of Normandy. Biographical profiles of each general are printed on the inside lid of the tin.

#109 A – General Charles de Gaulle
#109 B – General Bernard L. Montgomery
#109 C – Lt. General Omar N. Bradley
#109 D – General Dwight D. Eisenhower

#110 – 1995 VALENTINE Collection
Each high-polish chrome lighter features Valentine graphics and a "Be My Valentine" message. (Reminder: "ZOR No." stands for "Zippo's Own Reference Number")
#110 A – "Heart and 2 Cupids model" (ZOR No. 291)
#110 B – "Heart and Dove model" (ZOR No. 292)
#110 C – "White Hearts model" (ZOR No. 293)
#110 D – "Red and Pink Hearts model" (ZOR No. 294)

#111 – 1995 Zippo's 1995 Limited Edition Collectible…MYSTERIES OF THE FOREST
The 1995 Limited Edition (29, 325 lighters), Mysteries of the Forest (ZOR #MF-4), is the fourth in the annual "Lighter of the Year" series of collectibles from Zippo. The four-lighter set, when displayed in its collectors tin, completes the original Mysteries of the Forest artwork. The single lighter, Jaguar and Cub at Turtle Falls, is a companion piece to the set. These collectibles mark several firsts for Zippo. This is the first time the images on the lighters in a set combine to present a larger picture. It is the first Limited Edition to use Zippo's new Technigraphic imprinting, which provides photographic reproduction quality. And for the first time, there is a direct graphic connection between the set and the single lighter. As is customary with Zippo's Limited Editions, these collectibles are available only during the year of issue.
#111 A – Mystery #1
#111 B – Mystery #2
#111 C – Mystery #3
#111 D – Mystery #4
#111 E – Jaguar and Cub at Turtle Falls (Companion to the Four Mysteries Set)

#112 – 1995 18k GOLD

Zippo produced 18k gold models in lieu of their traditional 14k gold models.
#112 A – 18k Gold Regular
#112 B – 18k Gold Slim

#113 – 1995 ANTIQUE Finish Lighters

Zippo produced finishes in antique brass, copper, and silver-plate that were available plain or with vintage-look corner slashes.
#113 A – Reg. Antique Silver Plate (ZOR No. 121)
#113 B – Reg. Vintage Antique Silver Plate (ZOR No. 141)
#113 C – Reg. Antique Brass (ZOR No. 201)
#113 D – Reg. Vintage Antique Bras (ZOR No. 241)
#113 E – Reg. Antique Copper (ZOR No. 301)
#113 F – Reg. Vintage Antique Copper (ZOR No. 341)
#113 G – Antique Brass A6M Zero (ZOR No. 201VA-793)
#113 H – Ant. Brass Ju87 Stuka (ZOR No. 201VA-794)
#113 I – Ant. Brass F6F Hellcat (ZOR No. 201VA-795)

I was unable to acquire ZOR No.'s for #113 J-#113 N.

#113 J – Ant. Brass P-47 Thunderbolt
#113 K – Ant. Brass B-24 Liberator
#113 L – Ant. Brass P-51 Mustang
#113 M – Ant. Brass Curtiss P-40
#113 N – Ant. Brass F4U Corsair

#114 – 1995 MARBLE Collection
Laser engraved wooden display panel with American Classic logo. Each retailed for $43.95 by Zippo in 1995.
#114 A – White Marble
#114 B – Grey Marble
#114 C – Brown Marble
#144 D – Green Marble
#144 E – Red Marble
#144 F – Blue Marble

#115 – 1995 JIM BEAM Series
Available worldwide, this series was designed to mark the celebration of Jim Beam's 200th Anniversary. In 1795 Jacob Beam sold his first barrel of bourbon. That recipe was passed down through 200 years and six generations of his descendants. Today, Jim Beam continues to be made by Booker Noe, Jacob's great-great-great-grandson.

#116 – 1995 CHEVROLET Series
Available in USA only.
#116 A – Live the Legend (chrome)
#116 B – Genuine Chevy
#116 C – Chevy Truck
#116 D – Live the Legend (smoked)

#117 – 1995 POTPOURRI
Barrett-Smythe Collection, done only in midnight chrome. Offered in the Zippo salesman catalog.
#117 A – Chili PeppersXX
#117 B – Strawberry
#117 C – Tabby Cat
#117 D – Siamese Cat
#117 E – Peacock
#117 F – Brown Bear
#117 G – Raccoon

#118 – 1995 TRACK Series
#118 A – Daytona (Available in USA only)
#118 B – Indianapolis Motor Speedway (Available in USA and Europe)
#118 C – Talladega (Available in USA only)
#118 D – Darlington (Available in USA only)
#118 E – The Glen (Available in USA only)

#119 – 1995 DRIVER Series
Available in USA only. All retail for $24.95 except for #119 G.
#119 A – Rusty Wallace
#119 B – Kyle Petty
#119 C – Morgan Shepherd
#119 D – Bill Elliot
#119 E – Jeff Gordon
#119 F – Mark Martin
#119 G – Nascar Schedule (retails for $25.95)

#120 – 1995 CAR Series
Available in USA only. All retail for $24.95 by Zippo.
#120 A – Citgo #21
#120 B – Valvoline #6
#120 C – Miller Genuine Draft #2
#120 D – Dupont #24

#121 – 1995 EVENT Series
#121 A – Daytona 500 (Available in USA only and retails for $20.95 by Zippo)
#121 B – Indy 500 (Available in USA and Europe and retails for $19.95 by Zippo)
#121 C – Brickyard 400 (Available in USA and Europe and retails for $17.95 by Zippo)

#122 – 1995 RACING TRADEMARD Series
#122 A – NASCAR (Available in USA only and retails for $18.95 by Zippo)
#122 B – Winston Cup (Available in USA only and retails for $21.95 by Zippo)
#122 C – Winston Cup Drag (Available in USA only and retails for $17.95 by Zippo)

#123 – 1995 CENTENNIAL OLYMPIC Series
Available in USA only.
#123 A – 1996 Olympic Torch – silver plate (ZOR No 100A0 and retails for $24.95 by Zippo)
#123 B – 1996 Olympic Torch (ZOR No. M250A0 852 and retails for $24.95 by Zippo)
#123 C – 1996 Olympic Torch – brass (ZOR No 254BA0 851 and retails for $25.95 by Zippo)

#124 – 1995 WORLD WAR II Set
Available worldwide. Four piece set and "Victory Day" key ring retail for $99.95 by Zippo.
#124 A – Set of Four Lighters (ZOR No. ZR1-5)
   #124 A1 – Stars and Stripes Over Iwo Jima
   #124 A2 – Battle of the Bulge
   #124 A3 – Battle of Britain
   #124 A4 – U.S.S. Missouri

#125 – 1995 ANHEUSER-BUSCH Set
Available in USA only.
#125 A – Classic A with Eagle
#125 B – Budweiser Racing
#125 C – Ice
#125 D – Proud To Be Your Bud
#125 E – Busch
#125 F – Budweiser
#125 G – Bud Light
#125 H – Budweiser Clydesdale

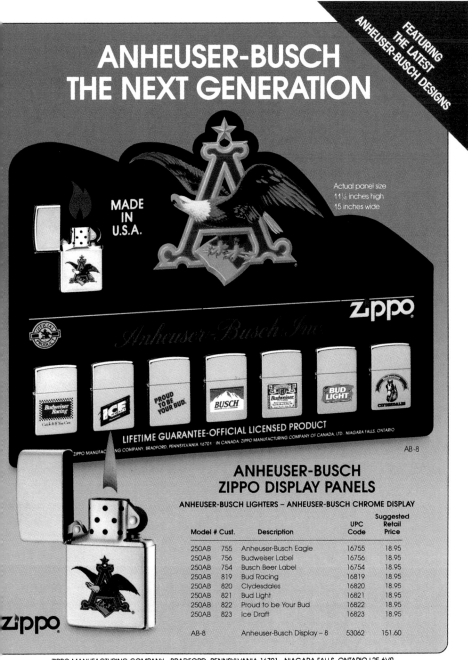

#126 – 1995 BARRETT-SMYTHE EMBLEM Set

#126 A – "Eagle" Brass Emblem (ZOR No. 254BBSB93. Retails for $23.95 by Zippo)

#126 B – "Eagle" Pewter Emblem (ZOR No. 250BS B89. Retails for $21.95 by Zippo)

#126 C – "Crocodile" Brass Emblem (ZOR No. 254BBS B75. Retails for $23.95 by Zippo)

#126 D – "Crocodile" Pewter Emblem (ZOR No. 250BS B74. Retails for $21.95 by Zippo)

#126 E – "Gorilla" Brass Emblem (ZOR No. 254BBS B94. Retails for $23.95 by Zippo)

#126 F – "Gorilla" Pewter Emblem (ZOR No. 250BS B90. Retails for $21.95 by Zippo)

#126 G – "Polar Bear" Brass Emblem (ZOR No. 254BBS B71. Retails for $23.95 by Zippo)

#126 H – "Polar Bear" Pewter Emblem (ZOR No. 250BS B70. Retails for $21.95 by Zippo)

#126 I – "Cobra" Brass Emblem (ZOR No. 254BBS B73. Retails for $23.95 by Zippo)

#126 J – "Cobra" Pewter Emblem (ZOR No. 250BS B72. Retails for $21.95 by Zippo)

#126 K – "Elephant" Brass Emblem (ZOR No. 254BBS B60. Retails for $23.95 by Zippo)

#126 L – "Elephant" Pewter Emblem (ZOR No. 250BS B59. Retails for $21.95 by Zippo)

#126 M – "Lion" Brass Emblem (ZOR No. 254BBS B62. Retails for $23.95 by Zippo)

#126 N – "Lion" Pewter Emblem (ZOR No. 250BBS B61. Retails for $21.95 by Zippo)

#127 – 1995 BARRETT-SMYTHE COMIC STRIP Series
Available worldwide. All retail for $17.95 by Zippo.
#127 A - ? (ZOR No. M200BS B120)
#127 B – POW! (ZOR No. M200BS B121)
#127 C – YIPE! (ZOR No. M200BS B122)

#128 – 1996 PINUP GIRLS Series: Pinup of the Year and the Four Seasons
The 1996 Limited Edition, Pinup Girls (ZOR #601), is the fifth in the annual "Lighter of the Year" series. This series made its debut in 1992 with the 60th Anniversary Lighter followed by the 1993 Varga girl, 1994 D-Day, 1995 Mysteries of the Forest, and 1996 Pinup Girls. As of September 1996, 108, 693 pinup units have been sold.
#128 A – Joan 'Pinup of the Year" (ZOR No. 601, retails for $29.95 by Zippo)
#128 B – Winter "Holly"

Numbers 128 B – 128 E (Set of four retails for $119.95)

#128 C – Spring "April"
#128 D – Summer "Sommer"
#128 E – Autumn "Ida Redd"

#129 – 1996 BARRETT-SMYTHE COMIC STRIP Series
All retail for $21.95 by Zippo.
#129 A - ? (ZOR No. M200BS B120)
#129 B – POW! (ZOR No. M200BS B121)
#129 C – YIPE! (ZOR No. M200BS B122)
#129 D - ! (ZOR No. M200BS B128)
#129 E - & (ZOR No. M200BS B129)
#129 F – ZAP (ZOR No. M200BS B130)

#130 – 1996 BARRETT-SMYTHE X-RAY Series
#130 A – Zippo X-Ray Pewter Emblem (ZOR No. 250BS B131. Retails for $25.95 by Zippo)
#130 B – Zippo X-Ray Brass Emblem (ZOR No. 254BBS B132. Retails for $28.95 by Zippo)

#131 – 1996 ZIPPO: WORLD WAR II, A Remembrance, Volume II Set
World War II has a special meaning for Zippo Manufacturing in that it was then that the Zippo lighter forged its reputation for reliability. This set contains four lighters with key holder that is packaged in a presentation case. This set commemorated both events and people of that pivotal conflict in world history. Four piece set and "Victory Day" key ring retail for $119.95 by Zippo.
#131 A – Limited Edition Set of Four Lighters (ZOR No. ZRII-5)
   #131 A1 – Battle of Midway
   #131 A2 – Flying Tigers
   #131 A3 – 3rd Army
   #131 A4 – Desert Fox

#132 – 1996 CHEVROLET Series
Ford or Chevy? The debate's been raging since drivers first chose their favorite. Zippo remains neutral, offering a series for each side. Sports car, pick-up, or family sedan, there's a quality American-made Zippo to match. Available in USA only.
#132 A – "Live the Legend Corvette" High Polish Chrome (ZOR No. 250CH 878. Retails for $21.95 by Zippo)
#132 B – "The Heartbeat of America" High Polish Chrome (ZOR No. 250CH 956. Retails for $19.95 by Zippo)
#132 C – "Chevy Trucks" High Polish Chrome (ZOR No.250CH 993. Retails for $19.95 by Zippo)
#132 D – "Genuine Chevrolet" Solid Brass (ZOR No 254CH 879. Retails for $22.95 by Zippo)

#133 – 1996 FORD Series
"The Best Never Rest." Available in USA only.
#133 A – "Ford" High Polish (ZOR No. 250F 957 Retails for $19.95 by Zippo)
#133 B – "Built Ford Tough" (ZOR No. 250F 958 Retails for $19.95 by Zippo)
#133 C – "Ford Trucks" (ZOR No. 250F 959. Retails for $19.95 by Zippo)
#133 D – "Ford Mustang" (ZOR No. 250F 960. Retails for $20.95 by Zippo)

#134 – 1996 JAMES BOND MOVIE POSTERS Series
#134 A – High Polish Chrome (ZOR No. 250BND 938. Retails for $31.95 by Zippo)
#134 B – High Polish Chrome (ZOR No. 250BND 939. Retails for $25.95 by Zippo)
#134 C – High Polish Chrome (ZOR No. 250BND 940. Retails for $29.95 by Zippo)
#134 D – High Polish Chrome (ZOR No. 250BND 941. Retails for $25.95 by Zippo)
#134 E – High Polish Chrome (ZOR No. 250BND 942. Retails for $29.95 by Zippo)
#134 F – High Polish Chrome (ZOR No. 250BND 943. Retails for $29.95 by Zippo)
#134 G – Solid Brass (ZOR No. 254BBND 945. Retails for $25.95 by Zippo)
#134 H – High Polish Chrome (ZOR No. 250BND 946. Retails for $23.95 by Zippo)

#135 – 1996 JAMES BOND GOLDENEYE Series
#135 A – Brass (ZOR No. 254BBND 937. Retails for $25.95 by Zippo)

#135 B – Chrome (ZOR No. 250BND 936. Retails for $23.95 by Zippo)

#135 C – Brass (ZOR No. 254BBND 936. Retails for $25.95 by Zippo)

#135 D – Chrome (ZOR No. 250BND 937. Retails for $23.95 by Zippo)

#135 E – Chrome (ZOR No. 250BND 935. Retails for $29.95 by Zippo)

#136 – 1996 TED LAPIDUS Series

#136 A – High Polish & Gold Plated (ZOR No. 250GTL 907. Retails for $33.95 by Zippo)

#136 B – High Polish & Gold Plated (ZOR No. 250GTL 908. Retails for $42.95 by Zippo)

#136 C – High Polish Chrome (ZOR No. 250TL 909. Retails for $24.95 by Zippo)

#136 D – High Polish Chrome (ZOR No. 250TL 910. Retails for $24.95 by Zippo)

#136 E – High Polish Chrome (ZOR No. 250TL 911. Retails for $23.95 by Zippo)

#136 F – High Polish Chrome (ZOR No. 250TL 912. Retails for $24.95 by Zippo)

#136 G – High Polish Chrome (ZOR No. 250TL 913. Retails for $24.95 by Zippo)

#136 H – High Polish & Gold Plated (ZOR No. 250GTL 914. Retails for $33.95 by Zippo)

#136 I – High Polish & Silver Plated (ZOR No. 100TL 916. Retails for $51.95 by Zippo)

#136 J – High Polish & Gold Plated (ZOR No. 250GTL 915. Retails for $33.95 by Zippo)

#137 – 1996 DRIVER Series

Winston Cup's hottest drivers and their rides are pictured on Motor Sports lighters. The driver portraits and car series are produced with Zippo's Technigraphic method for a photo finish. The 1996 NASCAR Winston Cup schedule can be found on lighter #137 G with full color NASCAR logo. Available in USA only. #'s A – F retails for $29.95 by Zippo.

#137 A – Rusty Wallace (ZOR No. 218N 830)

#137 B – Kyle Perry (ZOR No. 218N 831)

#137 C – Morgan Shepherd (ZOR No. 218N 832)

#137 D – Bill Elliot (ZOR No. 218N 833)

#137 E – Jeff Gordon (ZOR No. 218N 834)

#137 F – Mark Martin (ZOR No. 218N 835)

#137 G – 1996 NASCAR Schedule (ZOR No. 250N 992. Retails for $34.95 by Zippo)

#138 – 1996 EVENT Series

#138 A – Daytona 500 – High Polish Chrome; (ZOR No. 250N 948. Retails for $24.95 by Zippo) Available in USA only.

#138 B – Indy 500 – High Polish Chrome; (ZOR No. 250IMS 949. Retails for $22.95 by Zippo) Available in USA & Europe.

#138 C – Brickyard 400 – High Polish Chrome; (ZOR No. 250IMS 950. Retails for $21.95 by Zippo) Available in USA & Europe.

#139 – 1996 TRACK Series

#139 A – Daytona – (ZOR No. 250N 789. Retails for $23.95 by Zippo) Available in USA only.

#139 B – Indianapolis Motor Speedway (ZOR No. 250IMS 804. Retails for $21.95 by Zippo) Available in USA & Europe.

#140 – 1996 RACING TRADEMARK Series

Available in USA only.

#140 A – NASCAR (ZOR No. 250N 788. Retails for $22.95 by Zippo)

#140 B – Winston Cup Eagle (ZOR No. 250N 781. Retails for $25.95 by Zippo)

#141 – 1996 DALE EARNHARDT Series

Available in USA only.

#141 A – "Dale Earnhardt #3" High Polish Chrome (ZOR No. 250DE 961. Retails for $32.95 by Zippo)

#141 B – "Dale Earnhardt #3" Black Matte (ZOR No. 218DE 962. Retails for $23.95 by Zippo)

#141 C – "Intimidator with Checkered Flag" High Polish Chrome (ZOR No. 250DE 963. Retails for $32.95 by Zippo)

#141 D – "Seven Time Champion" (ZOR No. 250DE 964. Retails for $29.95 by Zippo)

#141 E – Brickyard 400 (ZOR No. 218N 930. Retails for $35.95 by Zippo)

#142 – 1996 CAR Series

Available in USA only. *All retail for $29.95 by Zippo.

#142 A – Citgo #21 (ZOR No. 218N 874)

#142 B – Valvoline #6 (ZOR No. 218N 875)

#142 C – Miller Genuine Draft #2 (ZOR No. 218N 876)

#142 D – Dupont #24 (ZOR No. 218N 877)

#142 E – Coors #42 (ZOR No. 218N 926)

#143 – 1996 ANHEUSER-BUSCH Series

Available in USA only. All retails for $22.95 by Zippo.

#143 A – Bud Ice – High Polish Chrome (ZOR No. 250AB 931)

#143 B – Red Wolf – High Polish Chrome (ZOR No. 250AB 932)

#143 C – Know When 2 Say When – High Polish Chrome (ZOR No. 250AB 933)

#143 D – Budweiser – High Polish Chrome (ZOR No. 250AB 995)

- #143 E – Bud Racing – High Polish Chrome (ZOR No. 250AB 996)
- #143 F – Bud Light – High Polish Chrome (ZOR No. 250AB 997)
- #143 G – Anheuser-Busch "Eagle" – High Polish Chrome (ZOR No. 250AB 755)
- #143 H – Budweiser Clydesdales – High Polish Chrome (ZOR No. 250AB 820)

#144 – 1996 CORONA Series
- #144 A – Corona – Black Matte (ZOR No. 218CR 979. Retails for $21.95 by Zippo)
- #144 B – Corona Extra – High Polish Chrome (ZOR No. 250CR 980. Retails for $20.95 by Zippo)
- #144 C – Corona Extra – Solid Brass (ZOR No. 254BCR 981. Retails for $22.95 by Zippo)
- #144 D – Corona "Bottle Cap" – High Polish Chrome (ZOR No. 250CR 982. Retails for $21.95 by Zippo)
- #144 E – Corona "Bottle Cap" – Black Matte (ZOR No. 218CR 983. Retails for $21.95 by Zippo)

#145 – 1996 RED DOG Series
- #145 A – Red Dog – Black Matte (ZOR No. 218RD 969. Retails for $21.95 by Zippo)
- #145 B – Red Dog/No logo – Black Matte (ZOR No. 218RD 970. Retails for $20.95 by Zippo)
- #145 C – Red Dog "y.o.u." – High Polish Chrome (ZOR No. 250RD 971. Retails for $20.95 by Zippo)
- #145 D – "The Dog's Red NOT The Beer! O.K." – High Polish Chrome (ZOR No. 250RD 972. Retails for $20.95 by Zippo)

#146 – 1996 COLT FIREARMS Series
- #146 A – Colt Python 357 Magnum – High Polish Chrome (ZOR No. 250C 984. Retails for $26.95 by Zippo)
- #146 B – Colt Anaconda 44 Magnum – High Polish Chrome (ZOR No. 250C 985. Retails for $26.95 by Zippo)
- #146 C – Colt King Cobra 357 Magnum – High Polish Chrome (ZOR No. 250C 986. Retails for $26.95 by Zippo)
- #146 D – Sam Colt (The Gun that Won the West) – High Polish Chrome (ZOR No. 250C 987. Retails for $23.95 by Zippo)
- #146 E – Colt "Horse" (ZOR No. 250C 988. Retails for $20.95 by Zippo)
- #146 F – Quality Makes It A Colt – High Polish Brass (ZOR No. 254BC 989. Retails for $25.95 by Zippo)

#147 – 1996 CAMEL Series
All retail for $27.95 and are offered directly by Zippo.

- #147 A – Antique Brass (ZOR No. 201FBCML 919)
- #147 B – Black Matte (ZOR No. 218CML 918)'
- #147 C – High Polish Chrome (ZOR No. 250CML 921)
- #147 D – Midnight Chrome (ZOR No. M250CML 920)

#148 – 1996 MILLER Series
- #148 A – "Lite" High Polish Chrome (ZOR No. 250MB 965. Retails for $21.95 by Zippo)
- #148 B – "Miller Genuine Draft" High Polish Chrome (ZOR No. 250MB 966. Retails for $22.95 by Zippo)
- #148 C – "Miller High Life" High Polish Chrome (ZOR No. 250MB 967. Retails for $21.95 by Zippo)
- #148 D – "Girl on Moon Motif" High Polish Chrome (ZOR No. 250MB 968. Retails for $27.95 by Zippo)

#149 – 1996 JIM BEAM Series
- #149 A – High Polish Chrome "Bottle Motif" (ZOR No. 250JB. Retails for $22.95 by Zippo)
- #149 B – High Polish Brass (ZOR No. 254BJB. Retails for $25.95 by Zippo)
- #149 C – High Polish Chrome (ZOR No. 250JB 928 Retails for $26.95 by Zippo)
- #149 D – High Polish Brass (ZOR No. 254BJB 929 Retails for $29.95 by Zippo)

#150 – 1996 BARRETT-SMYTHE THICK LIGHTER Series
- #150 A – Carpe Diem (ZOR No. 254BBS B149)
- #150 B – Champagne (ZOR No. 254BBS B150)
- #150 C – Full House (ZOR No. 254BBS B151)

#151 – 1996 CIGAR STORE INDIAN LIGHTER Series
- #151 A – Solid Brass (ZOR No. 254BBS B134. Retails for $28.95 by Zippo)
- #151 A1 – High Polish Chrome (ZOR No. 250BS B138. Retails for $25.95 by Zippo)
- #151 B – High Polish Chrome (ZOR No. 250BS B140 Retails for $25.95 by Zippo)
- #151 B1 – Solid Brass (ZOR No. 254BBS B136 Retails for $28.95 by Zippo)
- #151 C – Solid Brass (ZOR No. 254BBS B133. Retails for $28.95 by Zippo)
- #151 C1 – High Polish Chrome (ZOR No. 250BS B137. Retails for $25.95 by Zippo)
- #151 D – High Polish Chrome (ZOR No. 250BS B139 Retails for $25.95 by Zippo)
- #151 D1 – Solid Brass (ZOR No. 254BBS B135 Retails for $28.95 by Zippo)

#152 – 1996 TOLEDO BLACK MATTE Series
All retail for $40.95 by Zippo.
- #152 A – Mystical Lion (ZOR No. 520)

#152 B – Majestic Eagle (ZOR No. 522)
#152 C – Floral Portrait (ZOR No. 523)
#152 D – Eagle of Justice (ZOR No. 524)
#152 E – Medieval Knight (ZOR No. 525)
#152 F – Nature's Medley (ZOR No. 526)
#152 G – Bear's Fresh Fish (ZOR No. 527)
#152 H – Richard the Lion Hearted (ZOR No. 528)
#152 I – Portuguese Shield (ZOR No. 529)
#152 J – Fierce Guardian (ZOR No. 531)

#**153 – 1996 OLYMPIC GAMES CENTENNIAL Series**
#153 A – Torchmark – High Polish Brass (ZOR No. 254BAO 851. Retails for $30.95 by Zippo)
#153 B – Midnight Chrome (ZOR No. M250AO 852. Retails for $29.95 by Zippo)
#153 C – Torchmark – Silver Plated (ZOR No. M100AO 853. Retails for $29.95 by Zippo)
#153 D – Athens – Pewter (ZOR No. 250AO 973. Retails for $28.95 by Zippo)
#153 E – Athens – Brass (ZOR No. 254Bao 974. Retails for $30.95 by Zippo)
#153 F – "Atlanta 1996" Slim (ZOR No. 1654Bao 975. Retails for $24.95 by Zippo)
#153 G – Silver Plate with Gold Inlay in Laser Engraved Wood Box (ZOR No. OT-2. Retails for $109.95 by Zippo)
#153 H – Brass Torchmark Gift Set (ZOR No. OL-2. Retails for $39.95 by Zippo)
#153 I – 1996 Olympic Games Gift Set (1 of 10,000 sets – ZOR No. OL-7. Retails for $219.95 by Zippo)

#**154 – 1996 SOUTHWEST Collection**
All examples were produced in High Polish Chrome and retail for $32.95 by Zippo.
#154 A – Coyote Moon (ZOR No. 331)
#154 B – Death Valley (ZOR No. 332)
#154 C – Cactus Fire (ZOR No. 333)
#154 D – Buffalo Head (ZOR No. 334)
#154 E – Sundance (ZOR No. 335)
#154 F – Navajo (ZOR No. 336)
#154 G – Buffalo Stampede (ZOR No. 337)

#**155 – 1996 MIDNIGHT CHROME Set**
#155 A – Midnight Chrome-Brush Finish (ZOR No. M200. Retails for $15.95 by Zippo)
#155 B – Midnight Chrome-High Polish (ZOR No. M250. Retails for $16.95 by Zippo)
#155 C – Midnight Chrome with Zippo & Border (ZOR No. M250ZB. Retails for $16.95 by Zippo)
#155 D – Midnight Chrome with Zippo Emblem (ZOR No. M250ZE. Retails for $19.95 by Zippo)
#155 E – Midnight Chrome "Venetian" (ZOR No. M352. Retails for $17.95 by Zippo)
#155 F – Midnight Chrome "Tiger" (ZOR No. M200BS B28. Retails for $29.95 by Zippo)
#155 G – Midnight Chrome "Butterfly" (ZOR No. M200BS B42. Retails for $29.95 by Zippo)
#155 H – Midnight Chrome "Elephant" (ZOR No. M200BS B46. Retails for $24.95 by Zippo)
#155 I – Midnight Chrome "Tabby Cat" (ZOR No. M200BS B65. Retails for $25.95 by Zippo)
#155 J – Midnight Chrome "Peacock" (ZOR No. M200BS B67. Retails for $27.95 by Zippo)
#155 K – Midnight Chrome "Brown Bear" (ZOR No. M200BS B68. Retails for $25.95 by Zippo)

#**156 – 1996 PEWTER WILD WEST Series**
All retail for $28.95 by Zippo.
#156 A – Buffalo w/Necklace – Brass Emblem (ZOR No. 254BBS B141)
#156 B – Indian Chief – Brass Emblem (ZOR No. 254BBS B142)
#156 C – Cow Skull w/Necklace – Brass Emblem (ZOR No. 254BBS B143)
#156 D – Eagle over Mesa – Brass Emblem (ZOR No. 254BBS B144)

Table Models

#**157 – 1939 to 1940 – Barcroft Table Lighter**

1st Model

#157 B, $700-$1,000

This model was 4.5 inches tall and it had a "full insert." The 1939-40 model marked the beginning of Barcroft table lighters as well as table lighters in general. It was originally called the #10 Table Lighter De Luxe $7.50. The lighter had 14 holes in the chimney and a "one step" base, therefore it is commonly called the "One Step Table Model" between collectors. The fluid reservoir, located in the insert, extended to the bottom of the case.
The engraved model was called the #11 Table Model.
#157 A – Plain 1st Model
    (with an eight barrel hinge, composed of two four barrel hinges)
#157 B – Plain 1st Model (with a four barrel hinge)

#157 C, $2,500-$3,500

#159 B3, $1,500-$2,500

#159 B6, $1,500-$2,500

#157 C – Plain Brass 1st Model (with a four barrel hinge)

#158 – 1947 to 1949 – 2nd Model Barcroft Table Lighter

This model was 4.25 inches tall and had a full insert. Therefore, it was ¼ inch shorter than the 1st model. It was called the "#10 Table Lighter" in Zippo advertisements, whereas I found it listed as a "No. 10 Table Lighter DELUXE" on a June 30, 1949 Net Price List. The consumer price at that time was $10. It can be found having advertising, engraved initials, Sports scenes, etc. engraved on the face of the lighter. It is commonly called the "Two Step Tall Table Model" between collectors.

#158 A – Plain
#158 B – Town & Country Designs
#158 B1 – Trout
#158 B2 – Lily pond
#158 B3 – Horse
#158 B4 – Geese
#158 B5 – Setter
#158 B6 – Sloop
#158 B7 – Pheasant
#158 B8 – Duck
#158 C – Generic Advertiser
#158 D – Sports Series Motifs (line drawn)
#158 D1 – Tennis Racket
#158 E – Box and Instructions for 2nd Model Barcroft (I have seen 1st Model Barcrofts in the same box.)

#158 C, $200-$250

#158 D1, $600-$800

#158 E, $500-$700 (for empty box with paperwork)

#159 – 1949 to 1954 – 3rd Model Barcroft Table Lighter

This model is 3-1/4 inches tall and still has a full insert. It was called the "DELUXE Desk or Table Model" by Zippo, although I've seen it listed as a "No. 10 DELUXE Table Lighter" on a Confidential Price List. It is commonly called the "Short, Full Insert Table Model" by collectors. According to the Nov. 1, 1951 price list the consumer price was $10 while the retailer price was $6.

#159 A – Plain
#159 B – Town & Country Designs
#159 B1 – Trout
#158 B2 – Lily pond
#158 B3 – Horse
#158 B4 – Geese
#158 B5 – Setter
#158 B6 – Sloop
#158 B7 – Pheasant
#158 B8 – Duck
#158 B9 – Sailfish
#159 C – Generic Advertiser
#159 D1 – Zippo used a 3rd model to engrave this 1955-56 logo
#159 E – Sports Series Motifs (color-filled)
#159 F – Full Leather (test model)
#159 G – Ribbed Model (test model)
#159 H – Prototype Design
#159 I – Lil' Abner and Daisy Mae

#159 B1, $1,500-$2,500

#159 C, $250-$300. Unusual design (worth more)

#159 D1, $450-$600

#159 I, $250-$400

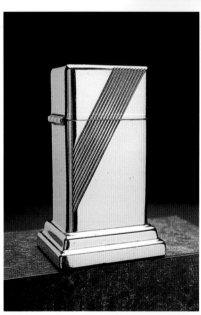

#159 G, Only one are known to exist. Valued @ $5,000-$7,000.

#160 – 1949 1st Model Lady Bradford (with no base)

This model was recalled the same year that it went into production. I was told a story that went something like this: A jewelry store that was carrying Zippos as part of its product line, had a customer that wanted to look at a 1st model Lady Bradford. While showing the lighter to the customer, the lighter tipped over and fell through the glass display case underneath it. Because of the incident, Zippo president, George Blaisdell recalled the item and made further modifications to the lighter by adding a wider base as support. It is this author's opinion, that Zippo drilled a hole in the bottom of the recalled lighter and used a screw and gasket to attach and seal the new support. In making this minor adjustment, Zippo could use the already existing case. If you pull the lighter wadding out of the fluid reservoir you are able to see the added screw attaching the new base to the bottom of the original lighter. If you remove the screw you can see where the hole was drilled through the word Zippo on the base of the lighter. It is this author's opinion that Blaisdell wouldn't have hidden the logo between the attached base and the bottom of

#159 H, Only three are known to exist. Valued @ $5,000-$7000

#160 A, 1949 1st Model (standing). $7,000-$9,000

#160 A, 1949 1st Model (bottom). $7,000-$9,000

the lighter nor would he have intentionally drilled a hole through the Zippo logo. Making this modification in this way was an extremely good idea. It was later marketed with the new modifications, the same year. This modified table lighter is commonly known as the 2nd Model Lady Bradford. Consequently, there are very few 1st Model Lady Bradfords known to exist, due to the recall.
#160 A – Plain

#161 – circa 1949-1954 2nd model Lady Bradford (with base)

This model was originally called the "No. 12 Lady Bradford" by Zippo. The 2nd Model Lady Bradford (with base), continued to be manufactured/sold, after the recall, until April 1, 1954, according to a "Confidential Price List." At that time the retailer's price was $6.27 whereas the consumer price was $11.50. The April 1, 1954 price list was the last one that the No. 12 Lady Bradford was found on.
#161 A – Plain

#162 B, $20,000-$25,000+

#162 – 1954 to 1979 – 4th Model Barcroft Table Lighter

This model was 3.25 inches tall but used the "same size insert" as the regular size pocket model. April 1, 1954, was the first time that the 4th Model Barcroft appeared on a Confidential Price List. At that time it was called the "No. 10 Barcroft Table Lighter" by Zippo. The distributor's price was $4.70 whereas the dealer price was $5.37. There was no listing for the consumer's price. I found the 4th Model on the May 1, 1955 Confidential Price List, even though it was already in production in 1954. Zippo didn't have all its old price lists. The May 1, 1955 Confidential Price List, was the last price list that I was able to find in Zippo's archives. It is possible to find motifs that were made using the Town and Country process on 4th model Barcrofts. The Town and Country process was still used as late as 1964. Town and Country illustrations can be found especially on prototypes and test models during that time frame.

#162 A – Plain

#162 B – Golfer Test Model (This is a Late 1950s – Early 1960s model.) This golfer was produced using the Town & Country process. This example is the same illustration that is found on the cover of the book.

#162 C – All Town and Country series motifs that were produced during this time frame. The illustrations Zippo offered depended on the year. See the Town and Country "Benchmark" Overview.

#162 D – Town and Country Advertisers

#162 E – Town and Country Ships

#162 F – Sports Series Motifs (line drawn)

#162 G – Sports Series (color-filled)

#162 H – Engraved Generic Advertiser

#162 I – Golf Ball (prototype)

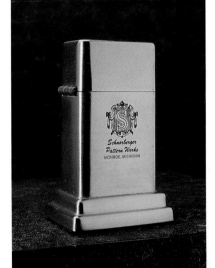

#162 H, $75-$100

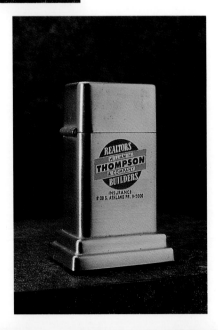

#162 H, $75-$100

#162 I, $1,500-2,000

#162 J – Porsche Imports Advertiser
#162 K – Alka-Seltzer
#162 L – Airstream (it was produced for a salesman's sample case)
#162 M – French's Mustard
#162 N – GM Allison Advertiser

#162 L, $175-$250. Salesman's sample.

#162 J, $125-$150

#162 M, $175-$250

#162 K, $175-$250

#162 N, $125-$150

#163 – 1960 MODERNE
The Moderne table lighter had a specially designed slim insert. It was produced in two finishes; chrome and black. It was marketed from 1960 to 1966. The insert had patent number 2517191®2940286 stamped on it.
#163 A – Chrome
#163 B – Black

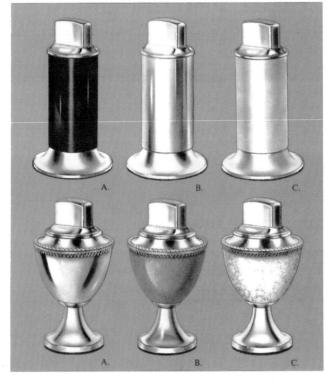

#163 B, $50-$60

#164 – 1960 CORINTHIAN

The Corinthian table lighter, like the Moderne, had specially designed slim inserts. The Corinthian was produced in three finishes; blue, pearl, and chrome. They were manufactured from 1960 to 1966. Both the Moderne and the Corinthian are the only two Zippo table models to ever use slim inserts.

#164 A – Blue
#164 B – Pearl
#164 C – Chrome
#164 D – Plaster of Paris prototype

Zippo would make plaster of Paris prototypes before making the finished product. This particular prototype has a prototype insert as well that fits perfectly into the base of the mold.

#164 D, Prototype, unique item, $3,500-$5,000

#164 A, $350-$500
#164 B, $250-$350
#164 C, $175-$250

#165 – 1979 HANDILITE

Zippo began production of the Handilite table lighter in 1979. This model is still being produced today. It is made of a regular pocket lighter with an attached, screwed on base.

#165 A – Generic Advertiser
#165 B – 7 UP
#165 C – 1980 Elephant
#165 D – 1980 Donkey

#165 A, $75-$100

#165 A, $75-$100

#165 C, $350-$500

#165 A, $75-$100

#165 B, $100-$150

#165 D, $350-$500

Zippo Accessories

#166 – 1962 RULE

Rulers can be found having advertising, Sports, and military motifs on them. They can also be found having other series illustrations on them as well.

#166 A – Generic Advertiser

#167 – 1964 POCKET KNIFE (without money clip)

The pocket knife was first introduced to Zippo's general line in 1964 after being test marketed for a number of years, according to long time employee Rudy Bickel.

#167 A – Town and Country Pheasant
#167 B – Sports Series Motifs
#167 C – Generic Advertiser
#167 D – Military Motifs
#167 E – Swiss Miss
#167 F – Zippo Fluid Can

#166, $15-$20

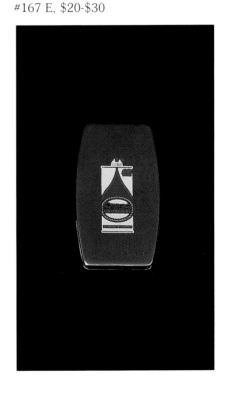

#167 E, $20-$30

#167 A, $500-$700. Only one known to exist in any Town & Country illustration.

#167 F, $150-$200. Extremely rare Zippo fluid can on a knife.

#168 B, $35-$50

#168 – 1971 MONEY CLIP/KNIFE Combo
#168 A – Generic Advertiser
#168 B – Jimmy Carter

Note: 1973 Key Holder has access #59. I inadvertently covered it at an earlier time in the text.

#169 – 1967 MAGNIFIERS
Magnifiers made their debut in 1976. Many of these had company logos as well as having "different series" motifs engraved on them.
#169 A – Generic Advertiser
#169 B – Bicentennial logo (This example happens to be two-sided.)

#170 – 1978 LETTER OPENER
Letter openers, like most Zippo products, can be found with advertising on them.
#170 A – Generic Advertiser

#171 – 1980 DESK SET
I didn't cover this at this time.

#172 – 1981 CUT-ABOUT KNIVES
#172 A – Generic Advertiser
#172 B – Military Logos

#173 – 1981 PILL BOX
#173 A – Disney Motifs
#173 B – Generic Advertiser

#169 B, $40-$60

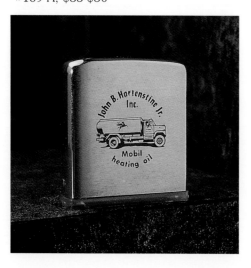

#169 A, $35-$50

#174 – FLINT DISPENSERS

Below:
Top row: c. 1938-early 1940s (#174 A, $125 & #174 B, $75), 1942-1949 (#174 C, $60), late 1940s (#174 D, $60), 1949-1962 (#174 E, $60).
Second row: Early 1950s (#174 F, $125), 1951-1953 (#174 G, $75), mid-1950s-c. 1964 (#174 H, $75), Customer imprinted: (#174 H1, $100), Service Zippo Flints, 1950s (#174, $30). **Third row:** 1962-c. 1967 (#174 J, $20), C. 1963-1978 (top: #174 K, $30; bottom #174 L1, $100).
Fourth row: 1960s-1970s (#174 M, $8), 1980s-current (#174 N, $1), 1978-1983 (top: #174 O, $10; bottom #174 R, $10), 1983-c. 1985 (top: #174 Q, $8; bottom #174 R, $8), 1985-current (top: #174 S, $1; bottom: #174 T, $1) Illustrated by Hiroshi Kito.

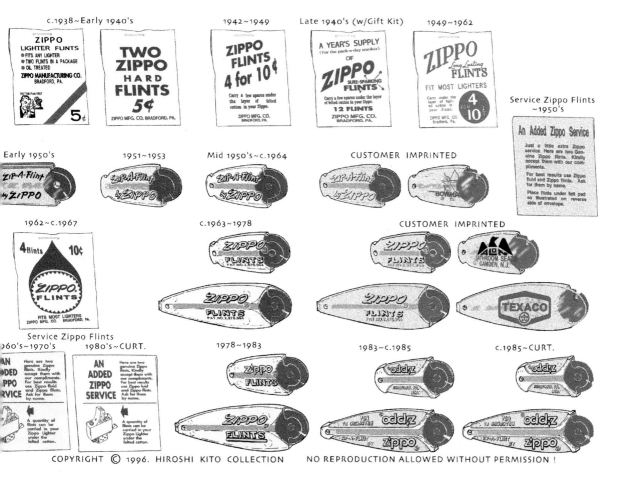

#174 U-Flint Dispenser Bottle

I am not sure when Zippo produced this dispenser, which is why they are not on the chart.

*Bulk dispensers were produced in the 1940s and 1950s. I have never seen any advertisements illustrating them.

#174 U, $300-$400

#174 H2, $115-$135

#174 L1, $30-$40

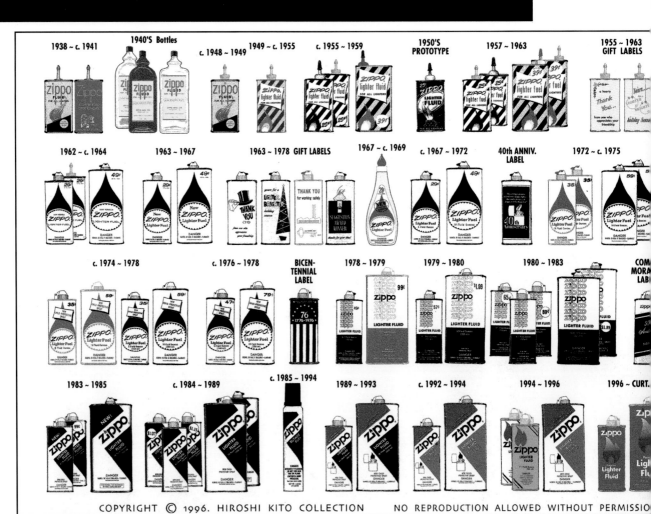

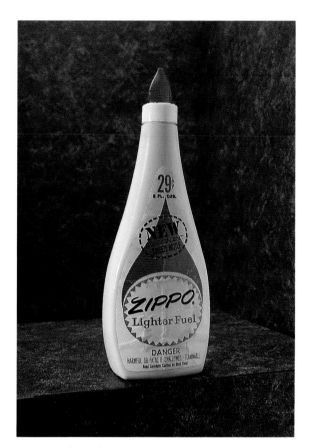

#175 A1, $400-$500

#175 – FLUID CAN

Top row: 1938-c. 1941 (#175 A, $500-$700; #175 B, $400-$600), 1940s bottles (#175 C, $400-$600; #175 D, $400-$600; #175 E, $400-$600) c. 1948-1949 (#175 F, $125-$175), 1949-c. 1955 (#175 G, $100-$150), c. 1955-1959 (#175 H, #175 I, and #175 J, $75-$125), 1950s prototype (#175 K, $275-$325), 1957-1963 (#175 L, #175 M, #175 N, and #175 O, $75-$125), 1955-1963 Gift Lables (#175 P and #175 Q, $100-$150. **Second row:** 1962-c. 1964 (#175 R, #175 S, and #175 T, $40-$60), 1963-1967 (#175 U and #175 V, $40-$60), 1963-1978 Gift Labels (#175 W, #175 X, #175 Y, and #175 Z, $75-$125), 1967-c. 1969 (#175 A1, $400-$500), c. 1967-1972 (#175 A2 and #175 A3, $15-$25), 40th Anniversary Label (#175 A4, $35-$45), 1972-c. 1975 (#175 A5, #175 A6, #175 A7 and #175 A8, $15-$25). **Third row:** c. 1974-1978 (#175 A9, #175 A10, #175 A11 and #175 A12, $15-$25), c. 1976-1978 (#175 A13 and #175 A14M, $15-$25), Bicentennial (#175 A15, $35-$45), 1978-1979 (#175 A16 and #175 A17, $8-$12), 1979-1980 (#175 A18 and #175 A19, $8-$12), 1980-1983 (#175 A20, #175 A21, #175 A22, #175 A23 and #175 A24, $8-$12), Commemorative Label, (#175 A25, $35-$45). **Bottom row:** 1983-1985 (#175 A26, #175 A27 and #175 A28, $4-$8), c. 1984-1989 (#175 A29, #175 A30, #175 A31, #175 A32 and #175 A33, $4-$8), c. 1985-1994 (#175 A34, $2-$4), 1989-1993 (#175 A35 and #175 A35, $1-$3), c. 1992-1994 (#175 A37 and #175 A38, $1-3), 1994-1996 (#175 A39, #175 A40, #175 A41, $1-$3), 1996-current (#175 A42 and #175 A43, $1-$3). Illustrated by Hiroshi Kito. *Please note: A 49 cent (1967-69 plastic lighter fluid bottle should be included in the lighter fluid cans list. It is not illustrated on the chart, but does exist. The chart was made before its known existence. I value both the 29 cent and 49 cent plastic fluid bottles the same. Both are extremely rare. I have given the 49 cent bottle access #175 A.

#176 – WICK DISPLAYS

Wick displays range in price from $10 to $200 depending on the age and condition of the display.
#176 A – 5¢ Asbestos Display
    #176 A   1 – Wrapper for the display
#176 B – 5¢ Asbestos Display

From this point on all the Zippo lighters and sundries are not in chronological order.

#176 A, $175-$225

#176 B, $175-$225

Military Lighters

#177

Zippo has manufactured lighters with ship motifs since World War II. The first military ship design was that of the U.S.S. Missouri with the surrender coin on the back. There are four known motif variations of this lighter (letters A-D) to the best of my knowledge. Zippo also used their famous electro-baked, Town & Country process to produce many different ship motifs until about 1960. Some of these are unbelievably beautiful, rare, and valuable.

#177 A – 1948-49 U.S.S. Missouri – This example has two coins; that of a ship on the front and a surrender coin on the back.

#177 B – 1949-50 Missouri – This example has an engraved ship on the front and the surrender coin on the back.

#177 C – 1949 Mid-Shipman Cruise Missouri – This variant has the ship engraved on the front with the words "Midshipman Cruise 1949," and the coin on the back. It also has General MacArthur's signature on it.

#177 D – 1952 Mid-Shipman Cruise – This example has the ship engraved on the front as well as saying "Midshipman Cruise 1952" with the coin on the back.

#177 E – 1983 Beirut-Lebanon Commemorative. This was a memorial lighter.

#176 A1, $50-$70

#177 A, obverse, $800-$1,000

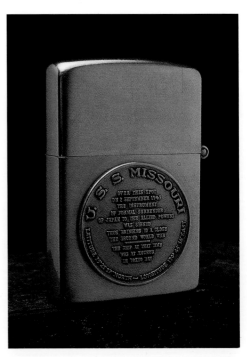

#177 A, reverse, $800-$1,000

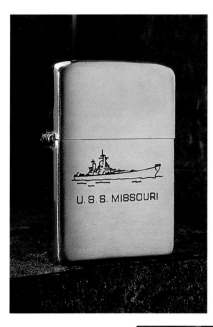

#177 B, obverse, $600-$800

#177 D, reverse

#177 B, reverse

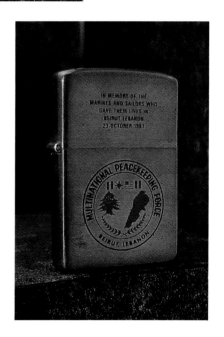

#177 E, obverse, $75-$125

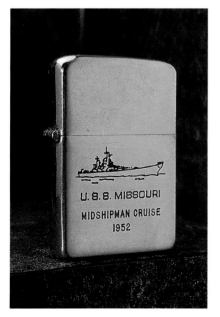

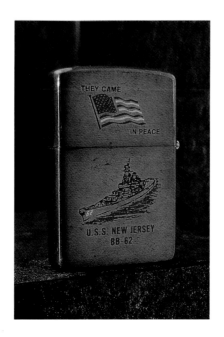

#177 D, obverse, $600-$800

#177 E, reverse

145

#177 F, $175-$250

#177 F – 2032695 pat. Aircraft Carriers (line drawn reg. Size)
    #177 F1 – 1949-50 U.S.S. Monterey
#177 G – 2032695 pat. Battle Ships (line drawn reg. Size)
#177 H – 2032695 pat. Destroyers (line drawn reg. Size)
#177 I – 2032695 pat. Support Vessels (line drawn reg. Size)
    #177 I – 1949-50 U.S.S. Timbalier AVP 54
#177 J – Town and Country Process Ships (reg. Size)
    #177 J1 – 1958 U.S.S. Forrestal
    #177 J2 – 1958 USS. Bayfield APA 33 (two sided)
    #177 J3 – 1961 USS Kenneth Baily DDR-713
    #177 J4 – 1964 U.S.S. New DD 818 (transitional T&C slim)
    #177 J5 – 1961 USS Aeolus ARC-3 slim
    #177 J6 – 1961 U.S.S. Blue DD 744
    #177 J7 – 1958 U.S.S. Greenwich Bay AVP 41 (The Green Witch)

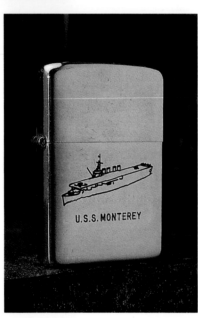

#177 F1, $175-$250

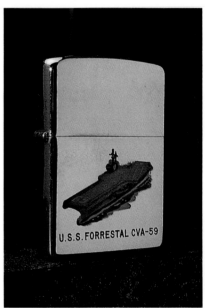

#177 J1, $500-$700

#177 I1, $175-$250

#177 J2, obverse, $800-$1,200. Worth considerably more since it is double-sided.

#177 J2, reverse

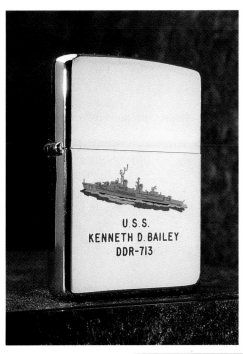
#177 J3, $500-$700

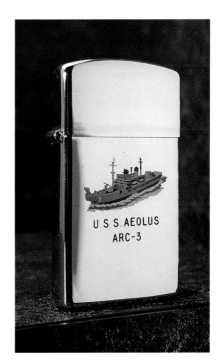
#177 J5, $300-$400

#177 J4, $300-$400

#177 J6, $500-$700

#177 J7, $800-$1,200. Awesome illustration.

#177 N1,
$200-$300

#177 K – 1953-56 Model Aircraft Carriers (reg. Size)
#177 L – 1956-56 Destroyers (reg. Size)
#177 M – 1953-56 Support Vessels (reg. Size)
#177 N – 1957-66 Aircraft Carriers (reg. Size)
    #177 N1 – 1966 U.S.S. Oriskany CVA 34
#177 O – 1957-66 Destroyers (reg. Size)
#177 O1 – 1966 U.S.S. Thomas J. Gary DER-326
#177 P – 1957-66 Support Vessels (reg. Size)
#177 Q – 1967-79 Destroyers (reg. Size)
    #177 Q1 – 1967 Lexington
#177 R – 1967-79 Destroyers (reg. Size)
#177 S – 1967-79 Support Vessels (reg. Size)
#177 T – 1980-90 Aircraft Carriers (reg. Size)
#177 U – 1980-90 Destroyers (reg. Size)
#177 V – 1980-90 Support Vessels (reg. Size)
#177 W – 1990-96 Aircraft Carriers (reg. Size)
#177 X – 1990-96 Destroyers (reg. Size)
#177 Y – 1990-96 Support Vessels (reg. Size)

Note, there are examples in each of the above categories (A-Y) that command substantially higher prices than others, such as those with cartoon characters depicting battle scenes or those motifs of famous ships. Also, regular size pocket lighters usually bring substantially more than slims but this is not always the case. In some cases slims bring substantially more than regulars due to their rarity and the desirability of the illustration. An example might be the 1959-1970 Sports models in which the slim test model version was that of a female and not a male. These are worth far more than their regular size counterparts. Also, it is this authors feeling that within a short period of time the illustration alone will determine the price independent of the fact that it is a slim or regular size Zippo. The example above already exemplifies this fact.

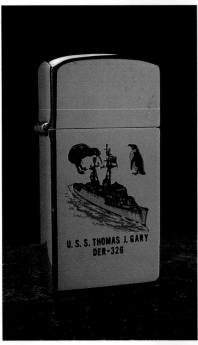

#177 O1, reverse,
$200-$300

#177 Q1, $200-$300

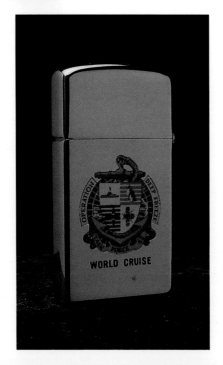

#177 O1, obverse,
$200-$300

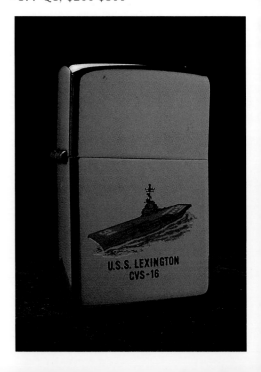

Desirable Advertisers

These are not chronological.

#178

The five most desirable post-World War II advertisers, from this author's perspective are: the Hertz emblem (with man that leaves and enters the car when the lid is up or down), GE Ultra-Vision TV with metal emblem and black paint, Kelvinator Refrigerator (with metal door handles that stand out from the case), Kennicott Copper Zippo (with solid copper case), and last but not least, the Alcoa Aluminum Zippo (with solid aluminum case). Many other "super advertisers" include those that were done using the Town and Country process, as well as those promoting Zippo products. The Grumman is a prime example. *See page 151.*

#178 A – 1985 Hertz Emblem ("Forefather" of the Trick lighters that Zippo produces today.)

#178 A, $1,000-$1,200

#178 A1 – 1975 Raggedy Ann and Tulip (similar to the Hertz Model)

#178 B – 1949-49 GE (Ultra-Vision TV (see advertisement sheets.)

#178 C – 1969 Kennecott Copper Advertiser (solid copper case)

    #178 C1 – 1971 Kennecott Copper Advertiser (solid copper case)

#178 D – Three Variations of Kelvinator Refrigerator

    #178 D1 – 1948-49 This model is painted in white enamel. It also has metal door handles on the lid and bottom half of the case.

    #178 D2 – 1948-49 This model has white enamel. It has only one metal door handle on the bottom of the case.

    #178 D3 – 1949-50 This model was never painted with white enamel, lacks the metal door handles, but has the metal Kelvinator Emblem

#178 E – 1959 Alcoa Aluminum Advertiser (with solid aluminum case which is anodized blue)

Above and left:
#178 A1,
$1,200-$1,500

#178 C1, $300-$500

#178 C, $400-$600

#178 F – 1955 Town and Country Grumman
  #178 F1 – 1966 Transitional T&C Grumman
#178 G – Advertisers Promoting Zippo and Zippo Products
  #178 G1 – 1971 29 cents Zippo Lighter Fluid Can with Flint Dispenser on the back
  #178 G2 – 1967 29 cents Zippo Lighter Fluid Bottle
  #178 G3 – 1970 Salesman's Sample
  #178 G4 – 1991 Zippo History
  #178 G5 – 1978 Zippo Zippo Zippo…
  #178 G6 – 1976 The Little Persuader
  #178 G7 – 1989 Zippo Advertisement
  #178 G8 – 1989 Zippo Advertisement
  #178 G9 – 1969 Blaisdell Portrait
  #179 G10 – 1992 Zippo Sailboat
  #179 G11 – 1975 47 cents Zippo Lighter Fluid Can (test sample)

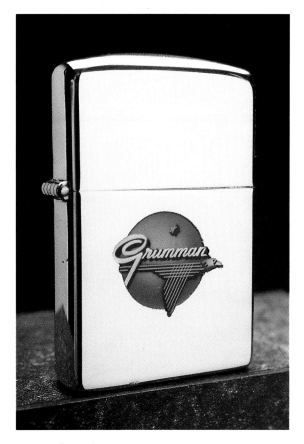

#178 F, $400-$600

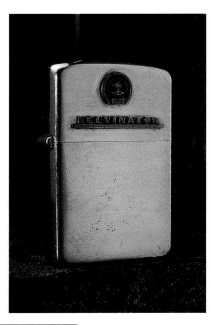

#178 D3, $250-$350

#178 E, $18,000-$22,000

#178 F1, $400-$600

Right and below:
#178 G1, $500-$700

#178 G3, $125-$150

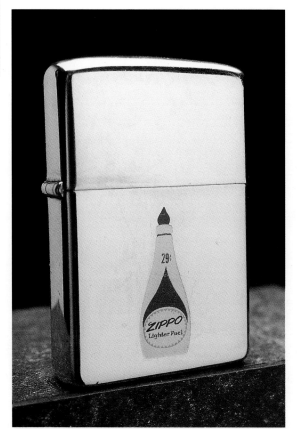

#178 G4, $50-$75

#178 G2, $1,000-$1,500

#178 G5, $75-$100

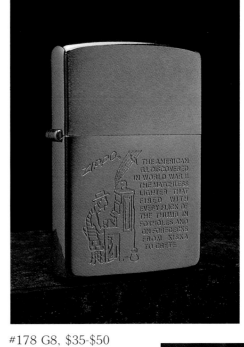

#178 G8, $35-$50

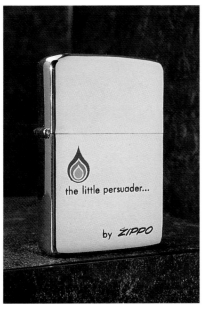

#178 G6, $85-$100

#178 G7, $35-$50

#178 G9, $3,000-$3,500

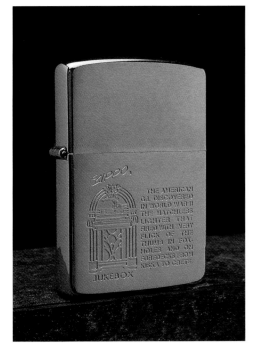

#178 G10, $50-$75

#178 G11, $400-$600

the advertising illustrations that Zippo has produced for a particular category. There are enough examples, I hope, to give you a representation of both rarity and price range. I will list a few examples, in each category, and their approximate value. With most any advertiser, it is the graphics as well as condition that determines the preponderance of it's value.

#179 – CAMEL

Camel appears to have started using Zippo lighters as an advertising medium during the 1970s. I am not aware of any examples that were produced before that time. Camel Zippo lighters are extremely desirable and collectible. Their desirability seems to have surpassed most other cigarette logos, possibly due to the prolific number of different illustrations that have turned up on the market in such a relatively short period of time. Also, the "Classic Camel" logo is easily recognized since it was used in the 1950s and 1960s on what are commonly called "flat advertisers," which were produced in Japan. The newer logo, using Joe Cool, is a motif that attracts both young and old alike. Circa 1991-92, Camel motif Zippo lighters could only be purchased with the C notes that one gets in a pack of Camel cigarettes. One C note certificate came in each cigarette pack. At that time it took 33 C note certificates to purchase one Camel logo Zippo. These C note certificates could be sent in and redeemed for whatever illustration Zippo had designed for Camel (R. J. Reynolds) at the time.

#179 A – 1992 Das Original
#179 B – 1989 Smooth Character (prototype)
#179 C – 1992 Camel Emblem
#179 D – 1992 Camel Emblem

*Author's note: Many of the early prototype and test model Camel Zippo lighters in mint condition trade in the $500 to $1000 range, depending on what is on them.

#180 – DISNEY

#180 A – 1991 Slim Mickey (Disney Logo)
#180 B – 1960 Mickey Emblem
#180 C – 1976 Walt Disney World Castle
#180 D – 1970 Walt Disney Productions Castle with D Initial
#180 E – 1973 Donald Duck
#180 F – 1972 Walt Disney Productions Mickey
#180 G – 1976 Walt Disney Productions Mickey Mouse (denim)
#180 H – 1976 Walt Disney Productions Mini Mouse (denim)
#180 I – 1987 Fort Wilderness Campground

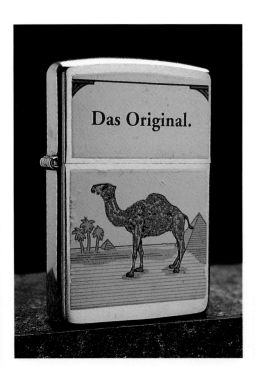

#179 A, $400-$500

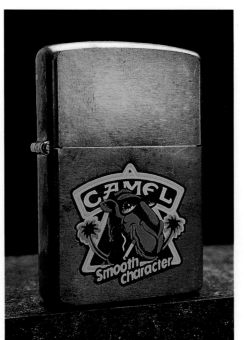

#179 B, $1,000-$1,200

This author, as well as any other, cannot begin to include all the rare and valuable advertising themes in one or even a number of books. Listed below are some collectible themes that I am most familiar with. These themes do not contain a comprehensive list of all

#179 C, $20-$30

#179 D, $20-$30

#180 J – 1981 Tokyo Disneyland (Brush Finish)
#180 K – 1981 Tokyo Disneyland (High Polish Finish)
#180 L – 1981 Tokyo Disneyland (slim)

#180 J, $1,500-$2,000

#180 K, $1,500-$2,000

#180 B, $800-$1,200. Extremely rare.

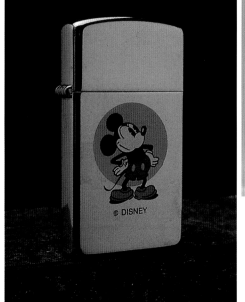

#180 L, $1,200-$1,500

Left:
#180 A, $300-$400

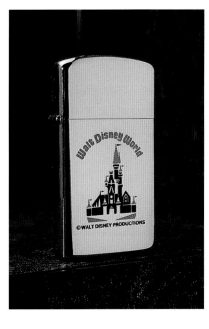

#180 C, $300-$400

155

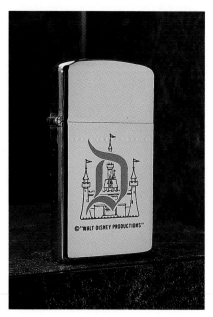

#180 D, $300-$400

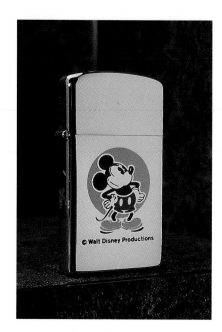

#180 F, $300-$400

#180 H, $800-$1,200

#180 E, $500-$700

#180 G, $800-$1,200

#180 I, $800-$1,200

#181 – FOOD AND SOFT DRINK
#181 A – 1968 Sunkist
#181 B – 1953 Pepsi Emblem
#181 C – 1990 Pepsi-Cola
#181 D – 1949-50 Coca-Cola

#182 – PETROLEUM
#182 A – 1982 Kendall Oil
#182 B – 1985 Kendall Motor Oil

#181 C, $50-$75

#181 A, $75-$125

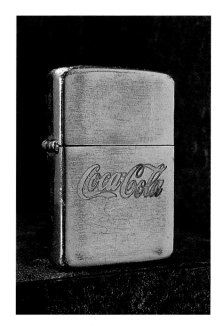

#181 D, $350-$500

#182 A, $350-$500

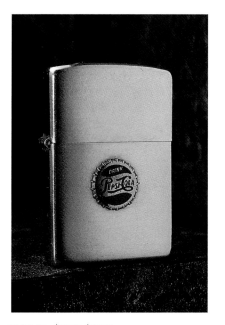

#181 B, $250-$350

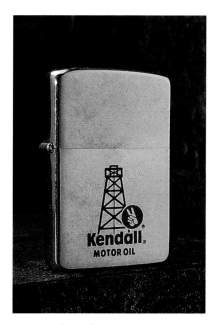

#182 B, $350-$500

#183 A, $250-$350

#183 – TRANSPORTATION
#183 A – 1948-49 Lyons Trucking
#183 B – 1960 Lyons Trucking (Town and Country process)

#184 – CIGARETTE
#184 A – 1958 L&M Filters
#184 B – 1968 Marlboro
#184 C – 1972 Marlboro
#184 D – 1986 Marlboro (engraved horse and rider)
#184 E – 1992 Marlboro (brass)
#184 F – 1949-50 Old Gold Cigarettes
#184 G – 1984 Newport Lights
#184 H – 1990 Marlboro (prototype)
#184 I – 1981 Lark

#184 A, $250-$300

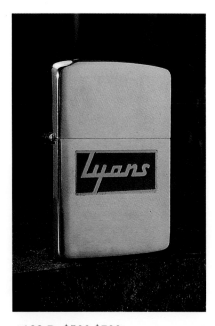

#183 B, $500-$700

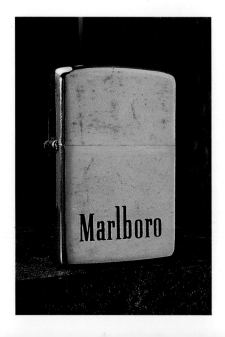

#184 B, $250-$350

#184 C, $250-$350

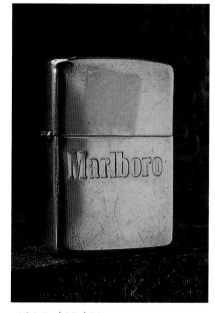

#184 E, $35-$50

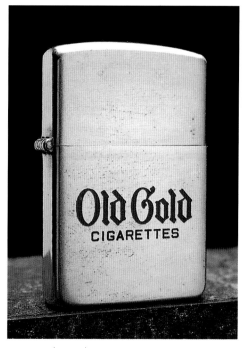

#184 F, $250-$300

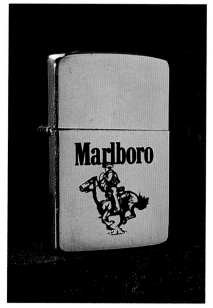

#184 D, $200-$300

#184 G, $100-$150

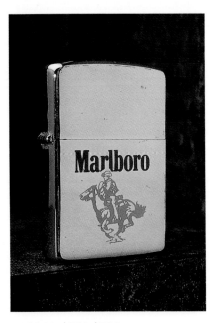

#184 H, $500-$700

#184 I, $150-$200

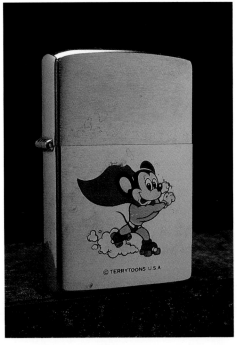

#185 A, $800-$1,200

#185 – CARTOON AND MOVIE CHARACTERS
#185 A – 1982 Mighty Mouse on Roller Skates (prototype)
#185 B – 1981 Dinky Duck (prototype)
#185 C – 1981 Deputy Dawg (prototype)
#185 D – 1982 Mighty Mouse on Skiis (prototype)
#185 E – 1985 Rocky IV (prototype)
#185 F – 1981 Popeye (prototype)

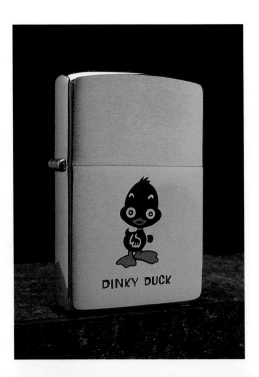

#185 C, $800-$1,200

#185 B, $800-$1,200

#185 D, $800-$1,200

#185 E, $800-$1,200

#185 F, $800-$1,200

#186 – BEER ADVERTISEMENTS
#186 A – 1982 Budman (prototype)

#187 – "Great Advertising Graphics in All Categories"
#187 A – 1973 Master Builders
#187 AA – 1959 Serenity Prayer (It has the smallest logo that I've ever seen)
#187 AB – 1980 Mt. St. Helens (prototype)
#187 AC1 – 1980 Mt. St. Helens (prototype)
#187 AC2 – 1980 Prototype (left); #187 AC3 – 1980 Production lighter (right)
#187 AD – 1962 DeVlieg Microbore (early silk-screen)
#187 AE – 1959 Cook Drilling Co. (T&C process)
#187 AF – 1961 Ed Black's Chevrolet (T&C process)
#187 AG – 1967 Two Ducks
#187 AH – 1975 Cameo (has very unique engraving process)
#187 AI – 1978 Pennzoil
#187 AJ – 1956 Sylvania

*Access numbers 185B, 185C, and 185F are prototypes due to lacking the copyright symbol which would have to appear on all production lighters.

#186 A, $800-$1,000

161

#187 A, $200-$300

#187 AC1, $250-$350

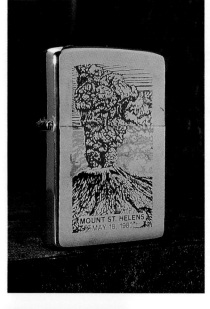

#187 AA, $200-$300. Zippo employee "Dale Hutton" test sample.

Left: #187 AC2, $250-$350; right: $187 AC3, $75-$100

#187 AB, $250-$350

#187 AD, $300-$400

#187 AE, $350-$500

#187 AG, $350-$500

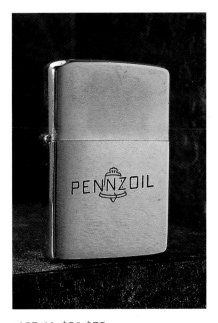

#187 AI, $50-$75

#187 AF, $800-$1,200

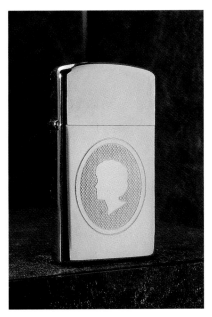

#187 AH, $750-$1,000. Prototype engraving process was used.

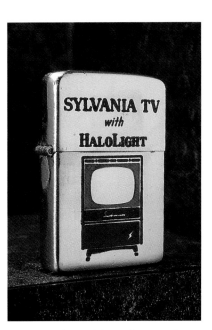

#187 AJ, $250-$350 (if in excellent to mint condition)

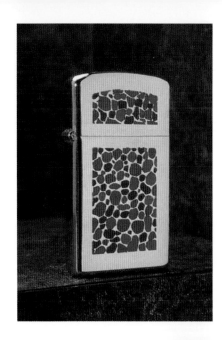

#187 AK – 1968 (prototype design)
#187 AL – 1989 Air Corps
#187 AM – 1991 Operation Desert Storm (test sample)
#187 AN – 1964 Elsie
#187 AO – Three Different 1977 Prototype Designs
#187 AP – 1978 Rods & Guns
#187 AQ – 1968 Pheasant (test sample). Remake of T&C Pheasant
#187 AR – 1970 Shreveport Times
#187 AS – 1990 Domino's Pizza

#187 AK, $350-$500

#187 AL, $35-$50

#187 AN, $175-$225

#187 AM, $250-$300

#187 AO, $300-$400 ea.

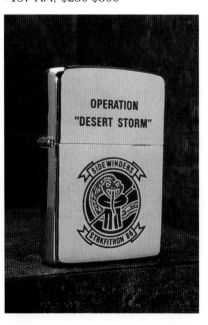

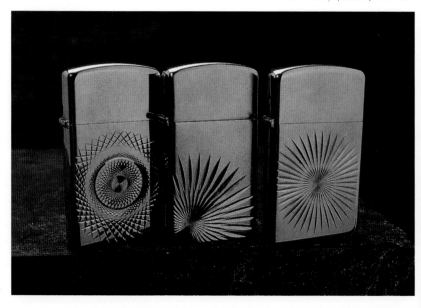

#187 AP, $75-$125   #187 AQ, $500-$700   #187 AR, $75-$125

Circa 1940-51 Zippo offered two different styles of Revellers, which could be engraved on a customers lighter if they sent it in with the fee. These illustrations are not exact.

#187 AT – 1940-48 Line Drawn Reveller. This variant has a rectangular border.
#187 AU – 1949-51 Line Drawn Reveller
#187 AV – 1989 Barcelona Olympics (Zippo never produced the illustration or finish of this prototype.)
#187 AW – 1985 "Amerigo Verpucci." (I believe that this prototype motif was produced for Italy. Production models lacked the blue sky in the background.)

Zippo car, 1947. See text on page 168.

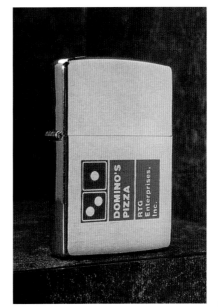

#187 AS, $10-$20

#187 AT

#187 AU

#187 AV, $125-$200

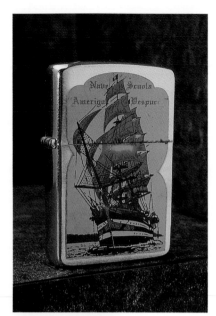

#187 AW, $150-$200

#187 B – 1996 Time "Marilyn Monroe" (prototype)
#187 C – 1969 Holiday Inn (salesman sample)
#187 D – 1940s Reddy Kilowatt
#187 E – 1975 Denim Wrangler
#187 F – 1992 Zippo Lighter Sketch
#187 G – 1969 Yellow Book Telephone (T&C process)
#187 H – 1962 John Deere (T&C process reg.)
#187 I – 1961 Bell The Best in Flowers (T&C process reg.)
#187 J – 1962 Angeletti Marble Co. Inc. (T&C process reg.)
#187 K – 1963 DelGrosso Spaghetti Sauce (T&C process reg.)

#187 B, $150-$200

#187 D, $350-$500

#187 C, $350-$500

#187 E, $100-$150

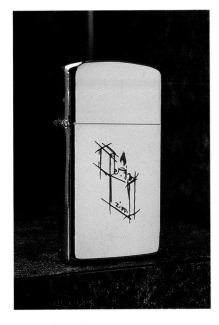

#187 F, $60-$70

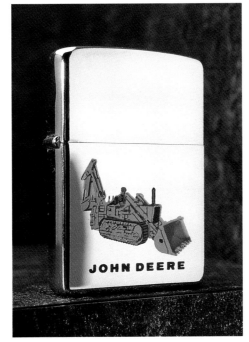

#187 H, $800-$1,200

#187 J, $500-$700

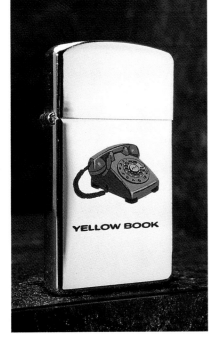

#187 G, $400-$600

#187 I, $500-$700

#187 K, $600-$800

#187 L, $600-$800

#187 M, $150-$250

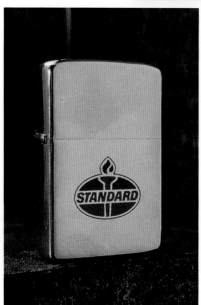

#187 N, $100-$150

#187 L – 1963 Morning Brew Coffee (T&C process reg.)
#187 M – 1967 Grumman (line drawn)
#187 N – 1975 Standard Oil
#187 O – 1995 Zippo Employee
#187 P – 1948-49 Zippo Car

Factoid: Concerning the Zippo Car (#187 P)

Date: August 25, 1986
Subject: Zippo Car

The Zippo car was on a 1946 Chrysler chassis. It was constructed by Gardner Display on Milwood Street in Pittsburgh, PA. When completed, Mr. Blaisdell hired Dick O'Day, (Dick was a car salesman in Bradford who had a dance band called "Dick O'Day and His Country Club Collegians") giving him a salary and expenses, to drive the Zippo car all around the USA. He called on wholesalers wherever he went. He also contracted the chief of police, and in many instances the Zippo car led

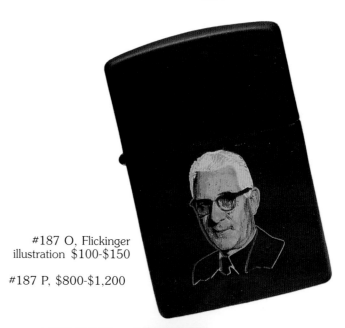

#187 O, Flickinger illustration $100-$150

#187 P, $800-$1,200

parades in various communities. Because it had an on board generator and loud speaker system it was great for crowd control in leading parades.

In 1950, when we started the first country wide District Manager representation, the Zippo car returned to Bradford and was used by the new District Manager's on a rotating basis from territory to territory.

In 1951 or early 1952, it wound up in Pittsburgh with Jim Pryor, the District Manager at that time. From the beginning, it was obvious that the chassis was not heavy enough to properly support the Zippo lighter built into the drivers compartment. Keeping tires on the vehicle was almost an impossibility.

Jim Pryor had a good friend, Henry Toohey, in Pittsburgh, who owned Toohey Motors, a Ford dealership. Toohey said he could solve the problem we had been having by putting the lighter display unit on a Ford truck chassis and use Mercury fenders, hood, trunk, etc. I cannot recall what the cost would be to make the transition, but Mr. Blaisdell gave Jim Pryor the okay to proceed. That was the end of the Zippo car.

I think Mr. Blaisdell was afraid Henry Toohey would call and say "your car is ready" and Henry Toohey was afraid Mr. Blaisdell would call and say "where is my car?" As a result, the vehicle, display unit, etc. disappeared through neglect and probably wound up in some junk yard.

In the late 1970s, he tried to resurrect the car or at least find out what happened to it, only to discover that Henry Toohey had died. His sons, Herb and George Toohey, had taken over the dealership and subsequently failed and went out of business.

Dictated by Bob Galey 8/22/86 (Former President and Sales Manager of Zippo Manufacturing Company)

#187 Q – 1967 Beagle
#187 R – 1962 John Deere Tractor (T&C process)
#187 S – 1968 Happy Hangover

#187 R, $800-$1,200

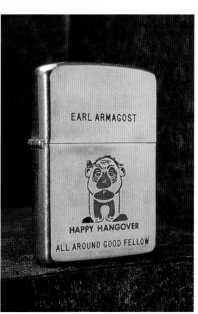

#187 S, obverse, $350-$500 (if in excellent to mint condition)

#187 Q, $200-$300

#187 S, reverse

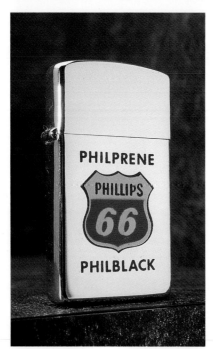

#187 T, $400-$600

#187 T – 1961 Phillips 66 (T&C process)
#187 U – 1994 World War II Caterpillar D7 lighter (Only 243 were manufactured in celebration of the 50th Anniversary of the CAT D7 in World War II.)
#187 V – 1960 U.S. Military Training Mission to Saudi Arabia
#187 W – 1949 ABC
#187 X – 1971 Deer
   #187 X1 – 1977 Deer (prototype winter variation)
#187 Y – 1951 Mallard Duck
#187 Z – 1980 U.S.A. Lake Placid (original model)

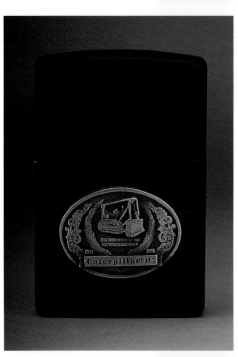

#187 U, $150-$200

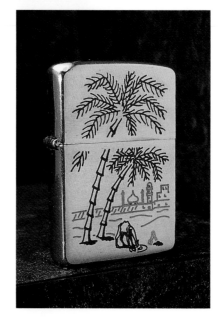

#187 V, obverse, $200-$300 (great graphic)

#187 W, $350-$500

#187 V, reverse

#187 X, $200-$300 (done like moon lander)

Miscellaneous Zippo Products

#188 "Great Design Changes"
#188 A – 1959 Prototype (Note that there is engraving on the sides and top of the lighter as well as the four faces of the lighter.)

#188 B – Prototype Hinge (Note that the date 7-14-62 was scratch engraved by the Research and Development Department. Also note that the middle hinge barrel is extremely long, like that of the 1938-41 sterling models.)

#188 C – Test model

#188 D – Test model

#187 X1, $400-$600 (done like moon lander)

#188 A, obverse, $400-$600

#188 A, reverse. Zippo engraved test model.

#187 Y, $350-$450

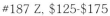

#187 Z, $125-$175

Above and right: #188 B, $750-$1,000. Unique hinge prototype.

#188 C, $500-$700

#188 D, $500-$700

#189 – ENGRAVING PLATES
#189 A – Bowman Products
#189 B – Desert Inn
#189 C – Cinch
#189 D – Carborundum
#189 E – Commercial Transport Inc.
#189 F – Comanche Supply Company
#!89 G – Sears Allstate Truck Tires

#189 B, $125-$175

#189 A, $125-$175

#189 C, $125-$175

#189 D, $125-$175

#189 E, $125-$175

#189 F, $125-$175

#189 G, $125-$175

#190 – ROSEART TALE Models

Zippo didn't actually produce these lighters although Zippo commissioned Phillip Rose, who had a mortuary business to produce them. Zippo supplied the inserts for them.

#190 A – Wood with Brass
#190 B – Wood with Chrome
#190 C – Black Marble and Brass
#190 D – Light Marble and Brass

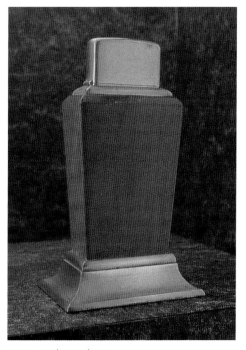

#190 A, $200-$275

173

#190 B, $200-$275

#190 C, $300-$500

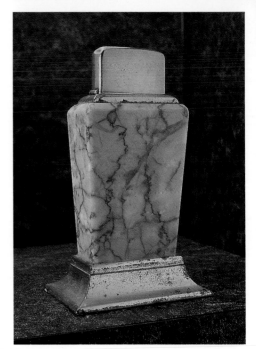

#190 D, $300-$500

#193 A – Tiffany Sterling Bamboo (slim)

#194 – "MINI" ZIPPO RENDITION
This was used in one of Zippo's advertisement campaigns.
#194 A – Card with "Mini" Lighter

#195 – ZIPPO CLOTHES LINE
#195 A – Clothes Line

#196 – ZIPPO ALARM CLOCK
#196 A – Alarm Clock

Below, left:
#191 A, $1,200-$1,500

Below right:
#192 A, $150-$250

#197 – ZIPPO ASHTRAY
This accessory was produced circa 1970. It has patent number 3,351,069. This accessory was only produced for about one year. Most examples have some type of advertising motif illustrated on them.

#191 – KENDALL/TABLE Model
The insert for a Moderne table model was used for the lighter.
#191 A – Kendall Can (prototype)

#192 – VEHICLE DASH BOARD LIGHTER HOLDER
#192 A – Zippo Lighter Holder

#193 – TIFFANY STERLING BAMBOO Lighter (has Zippo insert)

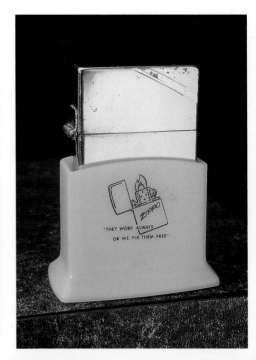

174

#193 A, $1,800-$2,500

#194 A, $75-$125

#195 A, $75-$100

#197 A – Submarine Advertisement

#198 ZIPPO ART WORK

This is an example of some original Zippo artwork. Zippo art work was very versatile. Zippo artists many times pulled old art work from their files and used it in the production of a new series, such as the Game Series. On the other hand, they might turn around and use a Game Series illustration in an advertising motif. I decided not to place a value on the art work.

#198 A – Heinz Ketchup

#198 B – Flying Turkey (Similar illustration was used in the Game Series.) This motif was produced 9/14/77.

#198 C – Grouse

#198 D – Trout (Similar illustration was used in the Game Series.) This motif was produced 12/8/80.

#198 E – Mallard Duck (Similar illustration was used in the Game Series.) This motif was produced 1/9/80.

#198 F – Soccer Player (Probably a prototype or test model design that was being considered in the Sport Series.)

#199 A

This is the 1st non-lighter Zippo product to my knowledge. Looks to made in the 1950s. It is a kaleidoscope. The box says "What's Behind Your Name?" Zippo. Valued at $5000-$7500

#196 A, $250-$350

#197 A, $150-$250 (depending on the illustration)

175

#198 A

#198 B

#198 C

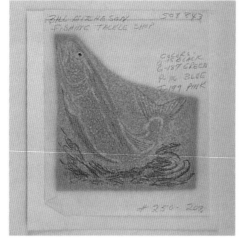

#198 D

#198 E

#198 F

176

1938 *Esquire* advertisement

A ZIPPO GLOSSARY

The Great American Lighter has produced its own words and phrases, a jargon particular to the Zippo historian and collector, that is essential to appreciating these pieces of Americana. Here is a sampling of "Zippoese," the language of lighters, to assist the reader in using this book.

Cam While "cam" refers to the metal piece that pushes the lid back when the lighter is open and being used, it is also the device that holds the cap down tight when the lighter is not in use. In both 1932/1933 tall models the cam was of a "hook shaped" design. (Note lighters #1 A and #2 A.) The cam was placed between two washers for smoother operation. Blaisdell switched to an extra large "figure eight" design in mid-1933 to accommodate the taller units. Some of the tall 1933 models were originally done with an extra large "figure eight" cam, probably to increase durability. When Blaisdell switched to making the ¼ inch shorter units, he also made the cam smaller, but of the same figure eight design, and he removed the two metal washers. Basically, the same figure eight design is still used today.

Cam Stop The function of the cam stop is to engage the cam, to hold the lid down when the lighter is not in use. The cam stop from 1932 through 1939 is of a U-shape design and was soldered inside the lid of the case. From 1940 to the present, the cam stop was actually an extension of the hinge which went up into the lid and curled around to supplant the function of the previously soldered cam stop.

Canned bottom This term refers to the contour of the bottom of the lighter. A lighter with a canned bottom is pressed in like the bottom of a soup can. Lighters of this style were first produced in 1946 with a 2032695 patent number on the bottom of the lighter. This shape is still produced today.

Flat bottom This refers to the contour of the bottom of the lighter. The bottom of the lighter can be either flat or slightly curved out. It is not pressed inward like the bottom of a soup can. Lighters of this style were only made up until 1945, except for replica examples that are still being produced. Lighters made from 1943-45 have an even more pronounced outward curve. These lighters were made of steel, for the most part, during World War II. The flat bottom shape was again used in the replica and commemorative lighters.

Flat Spring Different variations of flat springs have been produced from approximately 1936-37 to the present. A flat spring is that flat piece of metal that is riveted to the hole that the wick comes through. It extends across the top of the insert to the back, where the cam is located. It causes tension on the cam so that the cam is able to perform its two functions: that of holding the lid back when open, and that of holding the lid shut when not in use. A humped flat spring was used from 1937 to 1957. The straight flat spring made its debut in 1958 and is still used today.

Flint Tube The tube that contains the flint. It is accessed from the bottom of the insert.

Flint Wheel This is the round metal wheel that produces friction against the flint to produce the spark that lights the wick. They were horizontally cut until about 1946. The "Original 1946" was the last model to have a horizontally cut flint wheel.

Circa 1932-33: "Just Rite" Wheel
Circa 1934-42: "Alanklin" Wheel
Circa 1943-46: "Buckman" Wheel
Circa 1946-47: "Elge" Wheel (This wheel has criss-cross cut teeth.) **Factoid:** *Blaisdell spent $300,000 on this wheel.*

General Line Those lighters that Zippo sold through retail stores. General line lighters were sold as singles complete series sets, or as individual examples of important people, animals, places, events, company logos (such as Budweiser & Harley Davidson), etc. The Zippo sales representative would normally get brochures depicting lighters that were part of the general line stock.

Line drawing Also called "single line drawing," this is the style of engraving that Zippo used on its early advertising and sports series lighters. It is a simple outline sketch of the illustration. Most of the time it was done in one color, usually black. Other possible colors, depending on the exact time period, were red, black, blue, green, and orange. From 1939 to 1960 customers could send in their Zippo with $1.00 to personalize their lighter with a favorite Sports motif. Other line drawings such as initials, Reveller, etc., could be engraved on a Zippo lighter during this same period for $1.00. Dealers could order a display board and carry any combination of Sports motifs, depending on what the dealer thought would sell best in his geographic or ethnic location. Consumers had a choice of 19 "basic" Sports motifs between 1939 and 1951. (They are documented, as such, elsewhere in the book.)

Metal Emblem These are the hard metal emblems that Zippo put on lighters. These are different from metalliques in that they have a more rounded contour, whereas metalliques have a flat contour. These emblems illustrated fraternal organizations, branches of the military, etc. During the 1940s customers could send in their lighters and $1.00 and have it personalized with embossed designs of fraternal, club, or military insignia. Dealers could order a display board and carry any combination of metal insignias, depending on what the dealer thought would sell best in his or her geographic or ethnic location. This term might also be used to illustrate the metal emblems that a person might put on his or her own Zippo lighter, possibly during war or peace time, that would personalize the Zippo, and possibly make a statement.

Important: Keep in mind, that Zippo lighters made in the 1930s could also be sent in, after many years of use, and have one of three different styles of personalization applied to the surface: line engraved features (including signature facsimile, initials, sports motifs, etc.), metalliques (including monogram, Sports motifs, Reveller, Scotty group, etc.), or metal emblems (including organizational, club, and military insignias), for a minimal charge of $.50 to $1.00, depending on what services were provided by Zippo at the time. If a person wanted to spend the money, he or she could have had a number of "different types" of illustrations put on the same lighter for $.75 to $1.00, per illustration. For example, there is a 1938 model (access #9M) with a line drawn dog on one side, a line drawn bear on the lid of the opposite side, and a facsimile signature below the bear, on the bottom part of the case. It would have cost $3.00 for the illustrations on this lighter.

Metallique Flat metal emblems with color inlay that Zippo marketed from 1935 to the mid-1940s. These hair-thin metal slices are five one-thousandths of an inch thick and made of chrome plated brass. They were produced by the Probar Corporation of Orange, New Jersey (the company moved to Berkeley Heights, New Jersey, in 1937, at the height of metallique production.) Zippo used metalliques for the earliest Sports Series illustrations, as well as for advertising illustrations beginning with Kendall Oil. The fee in 1935 and early 1936 was $.50 to have the Scotty Group or Reveller motif applied to the surface of the lighter. Zippo raised the fee to $.75 in 1936. Lighters made before 1935 may be found with a metallique on them if they were sent in with

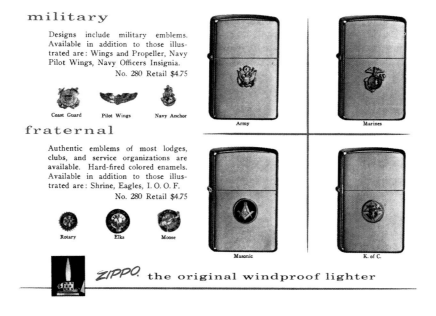

$.50 to $.75, depending on the fee at the time. Zippo also used metallique initials during this same period. Zippo charged $1.00 to have a person's metallique monogram applied. Monograms were been done in three styles (square, rectangular, or of a round design). Each monogram was color inlaid. The only color choices that were offered in 1936 were red, black, blue, green, and orange. About 1937, the customer had eight choices: black, red, blue, green, yellow, orange, purple, or white. Monograms were limited to four initials. Both services ended in the mid-1940s, for the most part and were replaced by the "Line Drawing" service. It has been reported, that some metallique company trade marks were still applied by 1950.

Model numbers Zippo used its own system of numbers and letters to identify the different models it produced over the years. Keep in mind that Zippo used the same model number for different variations of the same Sports illustration. They also used the same model number for a variation of the same style. For instance, Zippo used the #550 model number for both the full leather and leather wrap models. Also, the #10 model number referred to Zippo's Barcroft Table Model, whether it was a 1st Model or a 4th Model. (When describing one of "Zippo's Own Reference" numbers, this book uses the acronym "ZOR.")

Examples: (This is not a comprehensive list.)
No. 200 Brush Finish (all Sports lighters were done in the brush chrome finish with illustrations Nos. 175 A – 175 Q.
No. 180 Sports Series
No. 250 Bright Finish
No. 255 Facsimile Signature Lighter (brush finish case)
No. 275 TACH-A-LOOP; later called Loss-Proof with elastic lanyard
No. 275 S elastic lanyard with Sports design
No. 280 Military and Fraternal Organizations on brush chrome
No. 300 Personal Monogram Lighter; No. 300-1 horizontal without border; No. 300-2 vertical without border; No. 300-3 horizontal with border; No. 300-4 vertical with border (brush finish case)
No. 350 Engine Turned and chromium plated
No. 355 Signature Engraved on bright chrome lighter
No. 550 Leather
No. 750 Town and Country
No. 10 Table Lighter Deluxe
No. 15 Sterling Silver Plain Lighter
No. 20 Sterling Silver Engine Turned Lighter
No. 140 Plain Slim lighter in 14k Solid Gold
No. 150 Engine Turned Slim lighter in 14k Solid Gold
No. 165 14k Gold Plain Lighter
No. 175 14k Gold Engine Turned Lighter
No. 1500 Slim Plain Sterling Silver
No. 1700 Slim Engine Turned Sterling Silver
No. 1610 Slim High Polish Chrome
No. 1615 Slim High Polish Chrome Ribbon
No. 1621 Slim High Polish Chrome Bright Cut
No. 1625 Slim High Polish Diagonal Chrome
No. 1652 Slim Venetian Design

Outside hinge This refers to not only the hinge on the outside of the lighter but also the hinge plates. Technically, all Zippo lighters have the hinge on the outside of the case, but only those made from 1932-36 have the hinge plates, attached to the hinge, on the outside as well.

Plain model A model that may be plain (with no writing), have engraved initials, or have a facsimile signature done by Zippo Manufacturing Company.

Production Numbers Those collectors who are looking for production numbers are going to be disappointed. They, generally, do not exist.

Prototype This refers to the sample lighter that is made before the item is put into production. It is the early version of a design, made to help visualize the final product. According to several Zippo employees in the art department, usually five or fewer examples were made of a specific prototype unit. Again, prototypes that were produced for lighters in Zippo's general line and for an official series are most desirable and valuable. Prototypes that have major changes in the illustration, including background, colors used, or the engraving process, of course, are more desirable than a change in the surface of the lighter itself (such as having a brushed finish in contrast to a high polished chrome finish).

Important The illustrations for an advertising test sample or an advertising prototype are printed on the face of the lighter and sent to the consumer for final inspection before the end product was produced in 25 or more units (50 or more units today). The same was true for any "special order" illustration, even for the military. Those prototypes or test samples for a particular series were sometimes kept by Zippo or, over the years, sold to the employees.

Regulars This refers to the size of the case. This model could also be referred to as a "full size" or regular size case, in contrast to the slim model cases. Zippo's two most common lighters in this size, were their #200 and #250 models.

Series A series is a group of lighters depicting a particular theme. The first "official" series Zippo produced was the Sports series, which began circa 1937. It is still produced today. Zippo's instructional booklet lists many of the different series it produced over the years. Some examples in their booklet refer, it seems, more to an ongoing theme than to a specific set, such as the Sports, Space, Political, and NFL series, to name a few. Today, the term series refers more specifically to a "complete set" ranging, approximately, from two to ten lighters. Each lighter has an illustration which depicts a certain theme such as Vintage Trucks or Vintage Aircraft. Series sets of this type are usually produced from a few months to a few years, at most.

Signature Service If a customer sent his or her Zippo lighter in with a signature, Zippo promised to faithfully reproduce the signature on the lighter. The earliest ad extant for the facsimile signature is the 1938 Esquire advertisement. The signature service is still available today. The cost between 1938 and 1976 was $1.00. The 1991 salesman's catalog offers the service for $5.35 (retail).

Special Order A custom graphic (whether advertising or not) which the customer wants Zippo to put on a lighter. An individual had to order at least 25 units until about 1970. Today, an individual has to order a minimum of 50 units of a specific design, before Zippo will fill an order.

Test marketed During Zippo's early years, newly developed products were sent out with salesmen to see how well they would be accepted in the market place. If they sold well, Zippo would then carry the products as part of its general line. Today, it Zippo wants to "test market" a product, it often gives a sample of the product to its employees to try.

Important Even though a series or style, like the Political Series or Scrimshaw, was "officially marketed" in a certain year, Zippo test marketed a product for as long as one to two years before introduction. Zippo also used lighter cases left over from the previous year that were stamped with the previous code, before producing new cases with the new code the following year. Both reasons would account for "perceived" discrepancies concerning production years. Zippo would also go to its "throw away bins" to use the reverse side of cases to make practice illustrations for test models or prototypes. Zippo didn't always use these cases but often did. The "throw away bins" contained lighters that were unable to be sold due to some defect. The front of the case refers to the flat side facing the holder when the hinge is to the left of the case.

Test Sample (Test Models) Those individual designs that Zippo produced in limited numbers for a particular series that didn't get picked up for regular production are called test samples or test models. Those that are made for both the general line and for an official series are the most desirable and valuable. Test samples were also made illustrating customer trade marks and military graphics.

Town and Country Town and Country refers to the series that began circa 1947, with some examples being produced until approximately 1960. The original series consisted of eight lighters (duck, pheasant, geese, horse, trout, setter, lily pond, and sloop). In addition, the term also refers to a process which Zippo used during this same period to produce a high quality and a more expensive lighter. All the illustrations were done by hand. The illustrations were air brushed with acrylic enamel using different colors. Then the final product was electro-baked to increase durability. These are considered by most to be the most beautiful and desirable series to collect. They are very hard to find with 100% paint due to having no metal separating the colors. Because most people carried their lighters in their pockets along with coins, keys, etc. the metal-on-metal contact nearly always chipped the paint surface. As a result, mint quality Town and Country pieces are highly desirable. Zippo refers to the Town and Country series as "paint on paint." Advertising logos, military illustrations including ships, and the regular Town and Country series were done using this process until about 1960. Of course, line drawings were also produced during this time span for both advertising and military logos. (Wayne Edwards referred to the Town and Country series as Town and Country Sports lighters.)

Trench art A term taken from the World War I and World War II era, "trench art" refers to engravings, emblems, or other embellishments made to the lighters by soldiers during war. Thus the term has come to refer to the soldiers' customization, not Zippo's. Some of these changes particularly during the Vietnam War, were done by foreign engravers. Not all trench art was done by the hand of the owner.

DATING ZIPPO LIGHTERS

Regular Pocket Lighters

Below is a comprehensive list of the "Year Codes" of all REGULAR SIZE pocket models from 1932 to date. The codes are found on the bottom of the cases. Note, the size of the bottom of the cases are not necessarily proportional to each other or actual lighters. They were made on a "size needed" basis, to convey the thought. A magnet may be needed to help differentiate between years.

1932/33 Tall Model (without diagonal lines) – Access Number 1

This case is 2-7/16" tall, ¼" taller than 1934-1936 outside hinge plate motifs.

```
ZIPPO MFG. CO. BRADFORD, PA.
         Z I P P O
PAT. PENDING     MADE IN U.S.A.
```

1932/33 Tall Model (with diagonal lines) – Access Numbers 2 & 3

Case is still 2-7/16 inches tall during the 1st quarter of 1933. Mid-1933, Zippo reduced the case to 2-3/16 inches in height. True 1933's are 2-7/16 inches tall.

```
ZIPPO MFG. CO. BRADFORD, PA.
         Z I P P O
PAT. PENDING     MADE IN U.S.A.
```

1934 – Access Number 4

Case is 2-3/16 inches in height.

```
ZIPPO MFG. CO. BRADFORD, PA.
         Z I P P O
PAT. PENDING     MADE IN U.S.A.
```

1935 – Access Number 5

Case is 2-3/16 inches in height.

```
ZIPPO MFG. CO. BRADFORD, PA.
         Z I P P O
PAT. PENDING     MADE IN U.S.A.
```

1936 – Access Number 6

Case is 2-3/16 inches in height.

```
ZIPPO MFG. CO. BRADFORD, PA.
         Z I P P O
PAT. PENDING     MADE IN U.S.A.
```

This 1936 model must have an outside four barrel hinge and the above mentioned "PAT> PENDING" logo.

All true 1936's through 1941's have to have either a flat or slightly curved outward bottom and the 2032695 patent number. The 2032695 patent number was placed on the bottom logo, in mid-1936; *see access numbers 7-13*. The 1936-40 (TYPE 1) examples have flat bottom with sharp 90 degree angles. The 1938-41 (TYPE 2) variants have both corners that are rounded and bottoms that are slightly curved outward.

1936, TYPE 1, Access Number 7

```
ZIPPO MFG. CO. BRADFORD, PA.
         Z I P P O
PAT. 2032695     MADE IN U.S.A.
```

1937, TYPE 1 – Access Number 8

```
ZIPPO MFG. CO. BRADFORD, PA.
         Z I P P O
PAT. 2032695     MADE IN U.S.A.
```

1938, TYPE 1 – Access Number 9

```
ZIPPO MFG. CO. BRADFORD, PA.
          Z I P P O
PAT. 2032695      MADE IN U.S.A.
```

#1938, TYPE 2 – Access Number 9

```
ZIPPO MFG. CO. BRADFORD, PA.
          Z I P P O
PAT. 2032695      MADE IN U.S.A.
```

1939, TYPE 1 – Access Number 10

```
ZIPPO MFG. CO. BRADFORD, PA.
          Z I P P O
PAT. 2032695      MADE IN U.S.A.
```

1939, TYPE 2 – Access Number 10

```
ZIPPO MFG. CO. BRADFORD, PA.
          Z I P P O
PAT. 2032695      MADE IN U.S.A.
```

1940, TYPE 1 – Access Numbers 11 & 12

```
ZIPPO MFG. CO. BRADFORD, PA.
          Z I P P O
PAT. 2032695      MADE IN U.S.A.
```

1940, TYPE 2 – Access Numbers 11 & 12

```
ZIPPO MFG. CO. BRADFORD, PA.
          Z I P P O
PAT. 2032695      MADE IN U.S.A.
```

1941, TYPE 2 – Access Number 13

Zippo only fabricated the 1938-1940 TYPE 2 variant in 1941.

```
ZIPPO MFG. CO. BRADFORD, PA.
          Z I P P O
PAT. 2032695      MADE IN U.S.A.
```

All true 1943-1945 models will have the 2032695 PAT. NUMBER on a steel case. These do not have a canned bottom. The bottom of the case extends outward, even more profoundly than their 1938-1942 counterparts. You probably won't be able to see the patent numbers unless the black crackle paint has been worn off, since these were World War II models. *See access numbers 14 & 28.*

Note that some 1942 models have the **203695** patent number in lieu of the 2032695 patent number.

1942, TYPE 1 – Access Number 14

The 1942-1945 models have bottoms "similar" to the 1938-41 TYPE 2 variation.

```
ZIPPO MFG. CO. BRADFORD, PA.
          Z I P P O
PAT. 2032695      MADE IN U.S.A.
```

1942, TYPE 2 – Access Number 14

The patent number is missing the 2 in it. The letters are smaller and more rounded.

```
ZIPPO MFG. CO. BRADFORD, PA.
          Z I P P O
PAT. 203695      MADE IN U.S.A.
```

1942, TYPE 3 – Access Number 14

The word "Zippo" is larger than the other two variants (TYPE 1 & TYPE 2)

```
ZIPPO MFG. CO. BRADFORD, PA.
          Z I P P O
PAT. 2032695      MADE IN U.S.A.
```

1943

Ordinary type face but extra bold as far as the stamp. Zippo used the same dies through 1945.

```
ZIPPO MFG. CO. BRADFORD, PA.
          Z I P P O
PAT. 2032695      MADE IN U.S.A.
```

1944

```
ZIPPO MFG. CO. BRADFORD, PA.
          Z I P P O
PAT. 2032695      MADE IN U.S.A.
```

1945

```
ZIPPO MFG. CO. BRADFORD, PA.
          Z I P P O
PAT. 2032695      MADE IN U.S.A.
```

Chart of Cases and Inserts: 1946-1953

*Illustrated by Hiroshi Kito

	OUTER CASE				INSERT (Inside Mechanism)				
	MATERIAL	BARRELS	BOTTOM MARKINGS	LENGTH OF BOTTOM	MATERIALS	CHIMNEY & HOLES	TEETH OF FLINTWHELLS	MARKINGS	COPTT ENDS
1946 MODEL	NICKEL-SILVER (NO CHROME-PLATED)	PB#1 3-Barrel	PCM#1 PCM#2 No Chrome	PL#1 1-3/8"	NICKEL	PC#1 Old & 14 holes	PF#1 Horizontal Teeth	PIM#1 PAT. 2032695	PP# Cig.Fil-
1946-47 MODEL	NICKEL-SILVER WITH CHROME-PLATED	↓	PCM#1C PCM#2C with Chrome	↓	↓	↓	PF#2 Slash Teeth	↓	↓
1947 MODEL	↓	PB#1C PB#2 Shorter center barrel	PCM#3 ↓	1-11/32" PL#1 PL#2	↓	PC#2 New & 16 holes	PF#3	PIM#1C PIM#2 ZIPPO Logos	
1948-49 MODEL	BRASS WITH CHROME-PLATED	↓		1-11/32" ↓	↓	↓	↓	PIM#2	PP# Felt Pa
1949-1950 MODEL	↓	PB#3 5-Barrel							
1951 MODEL	BRASS WITH CHROME-PLATED	↓	PCM#3 PCM#4 PCM#5		NICKEL	PC#2	PF#3	PIM#2	
	STEEL WITH CHROME-PLATED				STAINLESS STEEL	PC#2C	↓	PIM#2S PIM#3S PIM#4S	
1952-1953 MODEL	STEEL WITH CHROME-PLATED		PCM#5A, #5B,#5C		STAINLESS STEEL ↓	↓		PIM#2 PIM#3	
1953 MODEL	↓		PCM#6					PIM#5 Full Stamp	

© Copyright, 1995 Hiroshi Kito. Used with permi

PB #1

Chrome-plated.

PB #2. Center barrel was shortened.

PCM #3. 3 barrel.

PCM #3. 5 barrel

PB #3. 5 barrel.

PCM #1

PCM #4

PCM #5

PCM #2

PCM #5-a. Regular.

PCM #1C. Chrome-plated.

PCM #5-b. Right pushed.

PCM #2C. Chrome-plated.

PCM #5-c. Hard pushed.

185

PCM #6

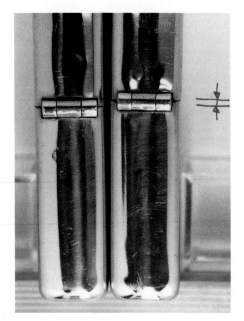

PL #1. A comparison of the longer and shorter bottoms from the hinge side. The one on the right is PL #1. The one of the left is PL #2.

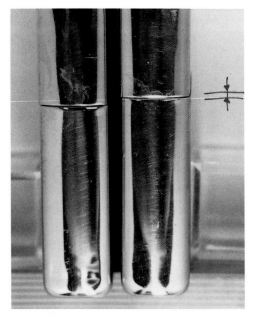

PL #2. A comparison of the longer and shorter bottoms from the other edge.

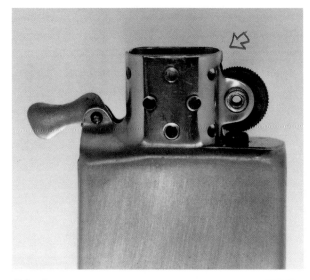

PC #1

PC #2

PC #2S

PF #1

PF #2

PF #3

PF #3S. Stainless steel.

PIM #1

PIM #1C

PIM #2

PIM #2S. Stainless steel.

PIM #3S

PIM #4S

PIM #2S

PIM #3S

PIM #5

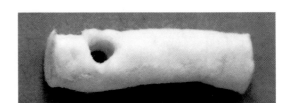
PP #1. Cigarette filter like cotton end.

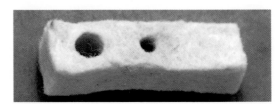
PP #2. Felt pad end.

Lighters manufactured from 1946 to date have a canned bottom with the exception of the replica lighters.

Note, when dating Zippos manufactured from 1946 to early 1953, with the 2032695 patent number on the bottom of the lighter, one needs to look carefully at the **exact placement** of the word "**ZIPPO**" in the logo in relation to the words "**MFG**" on the 1946-1951 models. See the photos on pages 185-188 for exact placement in relation to the entire logo.

1946

The 1946 model must have a 2032695 patent numbers as well as having "canned" bottom and a nickel/silver 3-barrel case.

TYPE 1

```
ZIPPO MFG. CO. BRADFORD, PA.
          Z I P P O
    PAT. 2032695   MADE IN U.S.A.
```

TYPE 2

The letters in the word "ZIPPO" are closer together.

```
ZIPPO MFG. CO. BRADFORD, PA.
          Z I P P O
    PAT. 2032695   MADE IN U.S.A.
```

1947

The 1947 model must have a chrome plated nickel/silver 3-barrel case.

TYPE 1

```
ZIPPO MFG. CO. BRADFORD, PA.
          Z I P P O
    PAT. 2032695   MADE IN U.S.A.
```

TYPE 2

Note, that the letters in the word "ZIPPO" are close together like the 1946 TYPE 2 variation.

```
ZIPPO MFG. CO. BRADFORD, PA.
          Z I P P O
    PAT. 2032695   MADE IN U.S.A.
```

1948-49

Note, the logo on the bottom is taller. It must also have a 3-barrel hinge.

```
ZIPPO MFG.CO.BRADFORD, PA.
          Z I P P O
    PAT. 2032695   MADE IN U.S.A.
```

1949-50

Note, the 1949-50 model has the exact same bottom markings as the 1949-49 model but it has a 5-barrel hinge in lieu of a 3-barrel hinge on a chrome plated nickel/silver case.

```
ZIPPO MFG.CO.BRADFORD, PA.
          Z I P P O
    PAT. 2032695   MADE IN U.S.A.
```

1951

Note, there were three different bottom logos on the 1951 model. All had 5-barrel hinges.

TYPE 1

This example must have a five-barrel hinge, on a steel case.

```
ZIPPO MFG.CO.BRADFORD, PA.
          Z I P P O
    PAT. 2032695   MADE IN U.S.A.
```

TYPE 2

Note, the spacing between the words MFG., CO., and BRADFORD in relationship to the word "ZIPPO."

```
ZIPPO MFG. CO. BRADFORD, PA.
          Z I P P O
    PAT. 2032695   MADE IN U.S.A.
```

TYPE 3

Note, the logo is shorter in height, and the words are close together, except for the word "ZIPPO" which is a little more spread out.

```
ZIPPO MFG.CO.BRADFORD,PA.
          Z I P P O
    PAT. 2032695   MADE IN U.S.A.
```

1952-53

Note, this model has a five-barrel chrome plated steel case. The 1951 TYPE 3 variant is exactly the same as this one.

Note: there were three bottom variations for the 1952-53 model. The logos were alike as far as placement, but were different as far as the depth of the strike that was made when stamping it. All had the "shorter and more compact" logo. All had five-barrel hinges. *See the photos (PCM #5-a, PCM #5-b, and PCM #5-c) on page 185 for exactness*

```
ZIPPO MFG.CO.BRADFORD,PA.
          Z I P P O
    PAT. 2032695   MADE IN U.S.A.
```

1953

Note, this model has the "full stamp" logo, on a chrome plated steel case, with a five-barrel hinge. Early 1953 models had the 2032695 patent number whereas later models had the 2517191 patent number with the large pat. pending logo that covered most of the bottom of the lighter. In 1953 Zippo started using the registered trademark subscript ®, as part of the bottom *logo*.

```
ZIPPO MFG.CO.BRADFORD,PA.
MADE IN   ZIPPO   U.S.A.
PAT. 2517191  ®  PAT.PEND.
```

1954-55

Note, it has a five-barrel hinge on a chrome-plated brass case.

```
ZIPPO MFG.CO.BRADFORD,PA.
MADE IN   ZIPPO   U.S.A.
PAT. 2517191  ®  PAT.PEND.
```

Changes in Zippo's Code System

According to Zippo's "Free Collectors Guide," the codes started circa 1958. I have always regarded this to be true, although upon close examination I have found it to be inaccurate. Most of the documentation that exists in Zippo's archives, such as the documentation by Wayne Edwards, was actually recorded 10 to 30 years after the fact by Edwards and other employees. This variable, probably accounts for most inaccuracies in Zippo's "Free Collectors Guide." The codes actually started in 1955.

Bottom of 1953-1955 case with no code.

1955 case with new Zippo logo and dot codes.

1956 case bottom.

1957 case bottom.

1958 case bottom.

1959 case bottom.

Zippo had been using its new logo in advertising since 1949, and in mid-1955 it changed the logo on all styles, including the regular model. Also, in 1949, the 1st model Lady Bradford had the new logo embossed on the bottom of the lighter, under the felt. The 1955 pocket model has four dots on the left and four dots on the right as well as having the small words, pat. pending to the right of the 2517191 patent number. The 1956 regular model has a code consisting of three dots on the left and four dots on the right side of the world Zippo, PAT 2517191 on the lower left, and the words PAT. PEND. (written in small letters), on the lower right side of the case. This is required if the lighter is a true 1956.

This was brought to my attention by a collector by the name of Sam Wood. Wood owns the Zippo lighter shown here with 1955 engraved by Zippo on the face of the anniversary lighter. This lighter has the above mentioned, 1955 code, on the bottom of the lighter.

Wood also owns the other two lighters illustrated here: National Homes and UMWA. Both have

commemorative logos with the year 1956 engraved on the cases. Both have the above mentioned 1956 identification code. One might argue that Zippo used older cases or some such explanation, but these three lighters have older engraved dates on newer cases, according to Zippo's current guide, not older cases with new logos.

Also, the one example is a 1956 Delegate Assembly lighter. The Delegate Assembly logo was only produced in 1956 for the Convention. It is unlikely that Zippo produced the 1956 logo on a 1959 case, when the stamp for the 1959 code wouldn't have been produced for three years. Fortunately, this knowledge only affects the 1955-59 codes. Therefore the 2517191 models, with the large patent pending logo were only manufactured from 1953 to mid-1955. The rest of the codes, it is assumed, are accurate.

The 25th Anniversary model is another example that supports these changes in Zippo's code system. The 25th Anniversary lighter was made by Zippo in 1957, yet according to the code on the bottom of the case, it was made in 1958. Again this is an example of an "older" date on a "new" case. It seems doubtful that at the end of 1957, Zippo got the idea for a 25th Anniversary Commemorative lighter, and produced all the Anniversary models in 1958. The code on the bottom of this lighter has four dots on the left and four on the right with the 2717191 patent number on the left (no patent pending logo). This a 1957 model, not a 1958. Zippo was awarded the 2517191 pat. in 1958. The 1958 model is different from the 1957 model in that it has three dots on the left and four on the right as well as having the PAT. 2517191 centered in the logo. Examples of the bottom logos from 1955 to 1959 are shown below.

If you look closely one can see the natural evolution that occurred during these years. If you check "anniversary models" with the date "Zippo" engraved on the face of the lighter, you will see that these bottom codes are accurate.

Note: In 1955 there were two bottom codes used. Zippo used the 2517191 patent with the "large" pat. pending logo as well as the dot logo shown here.

mid-1955 model

The word Zippo was stylized in script from 1955 to 1979. Note that the "PAT. PEND." logo is written in smaller letters.

1956 model

Note that one dot has been removed from the left side.

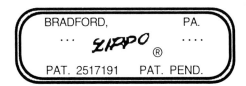

UMWA lighter.

National Homes lighter.

1957 model
Note that the dot on the left has returned, but the "PAT. PEND." logo has been removed.

1958 model
The 2517191 patent numbers are centered from 1958 to 1967.

1959 model
this is a true 1959 model, with 3 dots on the left and 4 dots on the right, but there is no PAT. PEND logo on a 1959 model. Also note that the only difference between a 1957 and a 1958 is the location of the patent number. Both models have four dots on each side.

The rest of the codes are as follows:

1960-1967 models have the 2517191 pat. # centered as well as having the dots or slashes depending on the year.

	LEFT	RIGHT
1960	....	....
1961	....	...

In 1961 Zippo began taking dots and slashes off the right side first instead of the left side as they had previously done, with the exception of the 1979 model.

1962	..	..
1963	..	.
1964	.	.
1964	.	.
1965	.	
1966	llll	llll
1967	llll	lll

Zippo must have ground the 2717191 patent number off the old dies in 1967 and then used the old dies to code the bottom of the new lighter cases. One can see a faint 2717191 patent number on the bottom of many 1967 & 1968 lighters. Some models made in 1967 have the 2717191 patent number while some don't.

1968	lll	lll
1969	lll	ll

Zippo made two changes on the bottom of the case mid-1969. Zippo used a new press machine in 1969 which caused the "canned" bottom of the lighter to be more dented in. Also, at this same time, Zippo changed the **Ƶ** logo on the word Zippo and gave the letter a "tail" hanging down on the right side." **Ƶ** " Therefore, there are two different logos for the bottom of a 1969 lighter.

1970	ll	ll
1971	ll	l
1972	l	l
1973	l	
1974	////	////
1975	////	///
1976	///	///
1977	///	//
1978	//	//
1979	/	//

1979 marked the last year that Zippo used the "Zorro" style **Ƶ** on the word Zippo. Also, for some reason the slashes were taken off on the left side first and not the right side.

1980	/	/

1980 was the first year of the stylized "**Zi**" in the word Zippo on the bottom logo. This bottom logo is commonly known as the Flaming "**i**" among collectors, although Zippo calls it their "Zi" logo. In addition, Zippo moved the words "Bradford, Pa." Below the word Zippo, on the bottom of the case. From 1980 to date Zippo has used many different bottoms logos, even for the same year (although their code system is still accurate).

1981	/	
1982	\\\\	\\\\

1982 marked the inception of the flat bottom Commemorative lighter in which Zippo designed to celebrate its 50th birthday. Commemorative lighters have been produced every year since 1982. Commemorative lighters will have both the 1932 date as well as the production year date on the bottom of the case.

1983	\\\\	\\\
1984	\\\	\\\

Circa 1984-85 Zippo devised and implemented a new variant logo for the bottom of the case although they still used the same codes. This new variation used a similar, but different version of the "pre-1969 Zorro stylized *Z* in the logo, as well as having a "flat bottom."

1985	\\\	\\
1986	\\	\\

Effective 7-1-86 the above system was replaced by a YEAR/LOT code. Year is noted with Roman Numerals whereas Letters designate LOT month (A=Jan., B=Feb. etc.)

1986	A to L II
1987	A to L III
1988	A to L IV
1989	A to L V
1990	A to L VI
1991	A to L VII
1992	A to L VIII
1993	A to L IX
1994	A to L X
1995	A to L XI
1996	A to L XII
1997	A to L XIII
1998	A to L XIV
1999	A to L XV
2000	A to L XVI

Slim Pocket Lighters

Below is a list of "Yearly Codes" for SLIM models.

	LEFT	RIGHT
1955	....	....

The insert must have both the wheel guard and modifications.

1956	....	....

The insert must have both the wheel guard and modifications.

1957	....	....

The insert lacks the wheel guard modification.

From 1958 to Date all the codes are accurate for slims in Zippo's Free Collectors Guide.

1958	...	...
1959	...	...
1960	..	...

For some reason the dots were taken off on the left side first and not the right side in 1960. The 1960 gold-filled models are an exception to this rule.

1961	..	..
1962	..	.
1963	.	.
1964	.	
1965		

Zippo put "no code" on the bottom of a 1965 slim, so that both the regular and slim size lighters would have the same code from then on.

1966	IIII	IIII
1967	IIII	III
1968	III	III
1969	III	II
1970	II	II
1971	II	I
1972	I	I
1973	I	
1974	////	////
1975	////	///
1976	///	///
1977	///	//
1978	//	//
1979	//	/
1980	/	/
1981	/	
1982	\\\\	\\\\
1983	\\\\	\\\
1984	\\\	\\\
1985	\\\	\\
1986	\\	\\

Effective July 1, 1986, the above system was replaced by a YEAR/LOT code. Year is noted with Roman Numerals whereas letters designate LOT month (A=Jan., B=Feb. etc.)

1986	A to L II
1987	A to L III
1988	A to L IV
1989	A to L V
1990	A to L VI
1991	A to L VII
1992	A to L VIII
1993	A to L IX
1994	A to L X
1995	A to L XI

1996 A to L XII
1997 A to L XIII
1998 A to L XIV
1999 A to L XV
2000 A to L XVI

Regular Size: 1953 to the Present

Please note that changes in inserts like cases did not usually begin January 1 of each year. Zippo made changes in the logo as well as changes in fabrication as they were needed.

1953-1956

Both the 1955 and 1956 inserts have the large PAT PEND. logo. This is the same insert that both the late 1953 and 1954 models have.

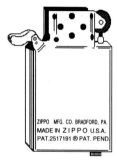

1957

This is actually both a 1957 and early 1958 insert, although I am going to call this a true 1957 insert. This insert lacks the PAT. PEND. logo. The 2517191 patent number is to the left side and not centered.

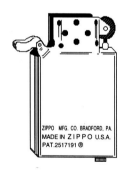

1958

This is actually a mid-1958 model although I am going to call this a true 1958 model. The patent number is centered on a true 1958 model but does not have the flints and fluid logo that you find on true 1959 models.

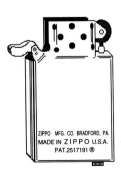

1959

Late in 1958 Zippo added the "Flints and Fluid" logo above the "Bradford/Pat. Number" logo, on the insert. I am going to call this a true 1959 although this model was also produced late in 1958.

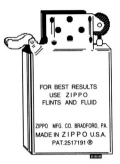

Late 1959-August 1, 1967

The patent number is not only centered but has the "Flints and Fluid" logo on the insert, like the 1959 model. The critical difference between a true 1959 and late 1959 insert can be determined by observing the placement of the logo. The logo on a true 1959 insert is written horizontally whereas the logo on a late model 1959 insert is written vertically. Zippo made two easily observable changes in the 1960s. The "hollow" brass flint wheel rivet was replaced with a "solid" rivet. The solid rivets appeared in 1961 but the hollow rivets were used until early 1963. Therefore, Zippo used the hollow rivet in the flint wheel from 1932 to circa 1963, although the rivet was made of different materials during this time frame. At times, the rivet was plated and at times it was not. Zippo also used red felt on the bottom of inserts manufactured mainly in the mid-1960s. Red felt has been documented on lighters as early as 1959 and as late as 1968.

Late 1967

Zippo removed the 2517191 pat. number after Aug. 1, 1967.

TYPE 1

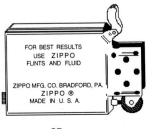

OR:

TYPE 2

1968-1975

The 1968-1975 models do not have any logo on the reverse side. This is a very important point for dating this insert. Also, in the early 1970s the straight "Cam Spring" was replaced with a widespread "Cam Spring."

TYPE 1

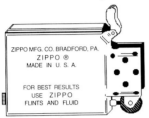

TYPE 2

1976-1981

The 1976-1981 models must have this logo on the reverse side as well as having the front logo:

Front

1982 to the Present

Lighters from dating from around 1982 to the present must have this logo on the reverse side. The order of the logo on the reverse side of a 1982 to date model is opposite that of the 1976-1981 logo. During that same time frame, Zippo began placing the same "Yearly Code" on the insert that was found on the bottom of the outer case, although its system was not very functional yet.

1982

1983

In 1983 there were two styles of inserts where the position of the logos were reversed.

TYPE 1

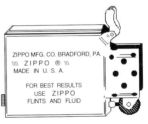

TYPE 2

1984

1985

1986

In 1986 Zippo used two styles of inserts with different logos, because they varied the "Yearly Codes" to a new form.

TYPE 1

TYPE 2

1987

Since 1987 all inserts have the same "A" to "L" markings as their outer case to identify the month in which they were produced. Note: Sometimes the year and month produced are almost unreadable in the insert.

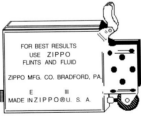

1988

1989

1990

1991

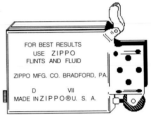

1992

1993

1994

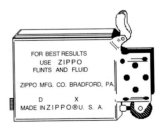

1995

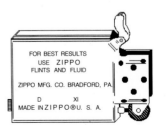

1996

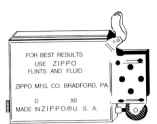

1997

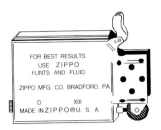

Slim Size: 1956 to the Present

According to Wayne Edwards, Zippo's Research and Development department started working on slim prototypes in 1955.

True 1956

As I stated earlier when discussing how to date cases, the insert of a 1956 slim lighter must have all the modifications that I listed in access number 43, as well as having the logo below and a slight dimpled chimney. The insert also has a "small hollow" wheel rivet as well as having the "early cam," which has an hour glass shape with sharp corners.

The exact placement and size of the logo on all slim inserts was very difficult to duplicate due to the width of the insert itself. Hopefully these insert drawings will prove beneficial. Also, if an insert appears to be a year off remember that Zippo would use all remaining stock from the previous year before producing new inserts with the new logo.

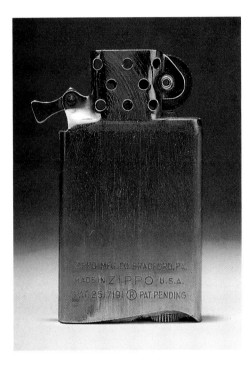

True 1957

This insert has the same logo as the 1956 insert but lacks the "wheel guard" modification that the 1956 insert has. To see the modifications that a true 1956 has, see access #43. This model still has a dimpled chimney, small hollow wheel rivet, early cam, early cam rest, and rear chimney view like the 1956 model. This insert can be found both having and not having the "PAT PENDING" logo.

True 1958

This insert variation still has the horizontal logo, although the "Flints and Fluid" logo has been added above the "Patent Number" logo. The words "patent pending" have been removed. The rear chimney view has also changed. This model still has the early cam, but the chimney has a more pronounced dimple due to a larger wheel rivet. One other important difference between this model and the previous one is that Zippo changed the cam rest. Some late model 1959 examples have a dented-in cam rivet.

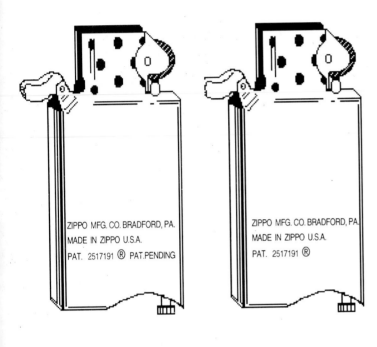

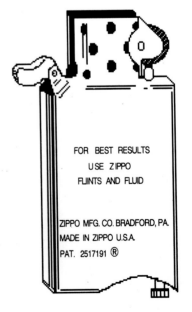

From this point on the cams, etc. are not illustrated in a technically accurate way, although the definitive differences of the "insert logos" can readily be seen in order to date the inserts. I will also try to include other definitive modifications in the text.

True 1959-1961

The logo runs the length of the insert from now on and the chimney on this model is no longer dimpled. The cam rivet, for the first time, is dented, whereas for the most part, it wasn't before. This model must have a "large hollow" wheel rivet.

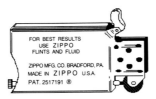

Overview of Slim Insert Modifications: 1961 to the Present

Circa 1964 Zippo replaced the "hollow" rivet in the flint wheel with a "solid" rivet on slim models. Zippo had been using the solid rivet on regulars since about 1961, although a few 1960 models have been documented as having the solid rivet. As for the Moderne and Corinthian slim table models, the hollow rivet was almost exclusively used except for those that were last in production, circa 1966. During the 1970s Zippo modified the wind holes on the chimney by making them smaller, as well as modifying the chimney's rear view.

1961-1967

Zippo used two types of inserts. Both variants had a "large solid" wheel rivet. From now on all models do.

TYPE 1

TYPE 2

1967-1976

This was a transitional year. Zippo used two types of inserts. Zippo dropped the 2517191 patent number from the logo, as of Aug. 1, 1967.

TYPE 1

TYPE 2

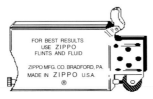

Zippo started putting notes of instruction on the reverse side of the insert in 1976.

1976-1979

Zippo reversed the order of the logo on this insert, compared to the previous variant.

TYPE 1

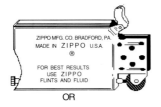

OR

TYPE 2

The bottom half of the logo on this variant has "smaller lettering" as well as the logo being more compact. Not the spacing between "FLINTS AND FLUID" and "ZIPPO MFG. CO. BRADFORD, PA."

1980

Zippo only produced this model in 1980.

Zippo reversed the order of the notes of instruction on the reverse side of the insert in 1981. This logo is still being used on the reverse side as of 1986.

1981-1982

The logo is no longer compact.

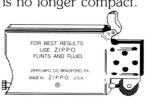

1983

Zippo produced two styles of insert logo.

TYPE 1

TYPE 2

1984

1985

1986

Zippo produced two different logos. The 1986 insert only has "G" to "L" markings representing the month of manufacture. *This is true of regular size inserts as well.

TYPE 1

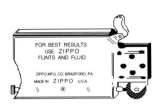

TYPE 2

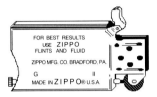

1987

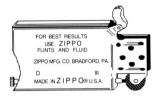

1988

1989

1990

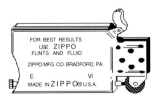

1991

1992

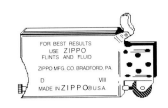

1993

1994

1995

1996

1997

During the 1990s many slims have been documented with an inverted logo. This is not uncommon or rare. Note where the position of the flint screw is when reading the insert right side up.

THE POORE PRICE GUIDE TO THE GREAT AMERICAN LIGHTER

Most people probably began collecting Zippo lighters for fun. It can rapidly evolve into an investment. There are two types of collector in the market: the casual collector who does it for fun and the investor who does it for investment and profit. There is room in the market for both the novice and the serious collector. The exciting thing is that there are Zippo lighters at the flea markets, garage sales, etc. that one might be able to purchase for as little as $5 and be able to sell for $300 to $1000 or add to one's collection, as I have done a number of times. One of my good friends, Bill Kim bought a "complete" 1932/1933 Zippo lighter for $1. Because vintage Zippo lighters are a true commodity and are bringing so much money, many who were casual collectors are now becoming investors to stay in the game. It gets harder and harder not to visualize a lighter that you paid $5000 for as an investment.

I have collected, sold, and traded thousands of Zippo lighters for the past fifteen years both in the United States and abroad, as well as attending most of the major cigarette lighter conventions. During this same time frame I have sold 1932/1933's as well as some of the more common Zippo lighters. My prices in this guide are based on this knowledge as well as talking to many of the more avid international collector/investors. These are prices that I would expect to pay in dealing with a knowledgeable collector. I'd like to make clear that Zippo Manufacturing Company has no interest or input into the prices of its vintage lighters in the collectible or antique market. Even if it wanted to (which it doesn't), Zippo could not put a price on the excitement of a collector who needs a single piece to start or complete a collection. That kind of zeal is priceless. Ultimately both the seller and buyer set the value of a lighter, which may be two to three times more than a price guide might indicate. When purchasing lighters please keep in mind that a collector will pay more for a lighter if it is one he needs to fill a vacancy in his collection and considerably less for one if he already has it. Also, prices tend to vary from coast to coast. I have taken all this into account when developing the guide to be fair to both the dealer and the collector. Always keep in mind that price guides are made to suggest a value, not to determine the price. Price guides are not fixed but meant to suggest a value.

When putting together a price guide, my friends told me that it was a "no win" situation. They said few would be happy with the prices established. There would be those who think the prices are too high and those who see them as too low. The differences between them comes down to varying perspectives, that is, whether or not the collector already owns the lighter in question. Those who own them want higher prices, while those who have yet to acquire them want to see a lower value...though only until they, too, become owners. I have valued the lighters according to what I would pay, have paid, or have sold them for when dealing with knowledgeable collectors. There will be those who act like they have never seen them sell that high or that low. Some haven't, but many have, including myself.

I suggest and encourage collectors to buy lighters that are in excellent to mint condition as well as aesthetically pleasing. These tend to hold their value and prove to be the better investment if you decide to liquidate your collection. Lighters that have been repaired in any way are worth a small fraction of their regular value, whatever that might be. It is better to have a loose hinge that is original than a repaired new hinge that is tight. What is a chipped rare vase worth compared to one that is not chipped? Of course, this does not mean that a 1932/1933 model for example, that has been repaired by Zippo lacks value. Not at all, but the analogy is still true. Most all 1932/1933's have been repaired by Zippo in some way as I stated earlier, but because they are so rare, they "march to their own drummer" in the area of value. Though both are valuable, the one that is not repaired in any way is worth considerably more. Overall, this price guide is meant to fuel your imagination as much as it is meant to inform our deal.

Ebay Prices

Ebay prices, for the most part, are lower than book prices. Almost, without exception, the best lighters are sold privately, apart from Ebay. Although, once in a while an awesome lighter does make it to Ebay's venue. Many Ebay buyers are not end buyers. They are buying lighters on Ebay to resell them in shops or to private collectors overseas. The end buyer realizes a lighters true worth and will ultimately pay the most for the item. I deal more with end buyers rather than speculators and these prices reflect that.

When it comes to rare Zippo lighters, the prices have risen drastically. Today, we know so much more about both rarity and value than we knew 8-10 years ago.

Pursue the 1932/33 Tall Model

When pursuing a 1932/1933 tall model please keep in mind that the vast majority that are extant have been repaired by Zippo. Also remember that if one of these lighters were sent to Zippo anytime until about 1980, the repair department would usually make what I am going to call a "complete" repair job. This meant that they would repair and replace all "resistance" parts (cam, coiled cam spring, hinge, and flint wheel) whether it needed it or not. This was done, first, because it was perceived as "good business." Zippo felt that they were doing their customers a service by replacing old parts with new. Over the years Zippo technology had improved. Better and more durable parts had been designed and fabricated. These new parts made the lighters more functional and durable. Throughout Zippo history this principle has been preeminent. They also made these repairs because they did not want a lighter that had just one thing repaired to come back in three months to have something else fixed. This would have created ill-will with the customer, while causing Zippo to accrue more expenses by having the lighter go through the repair department again.

In light of the marketing venue, caused by the Great Depression, most of the existing 1932/1933's ended up in the Bradford area. The few that survived were sent in to the Zippo plant there for repairs after prolonged use. If Zippo hadn't had such an unbelievable "Lifetime Warranty," far fewer would exist today in any condition. At that time Zippo lighter owners never gave a second thought to having their lighter repaired. It was considered one of the perks of living in Bradford.

Of course, today Zippo has a different repair policy, knowing how collectors feel about their prized possessions. If you are lucky enough to find a repaired 1932/1933 don't hesitate to buy it, if you can get the right price. It may well be the one chance in your lifetime to acquire this exclusive lighter. Few of the thousands and thousands of Zippo collectors have actually held or seen one in any condition. Good luck in your pursuit of the "Phantom Zippo."

Chart 1: Rarity Values

5 = Less than 15 known to exist (Extremely Rare)
4 = Between 15 and 50 known to exist (Very Rare)
3 = Between 50 and 100 known to exist (Special Order/Rare)
2 = Regular Production Lighter/Unknown # (Uncommon)
1 = Regular Production Lighter (Common)

*Lighters #1 A-#1 C must have totally original parts to be worth these prices.

Access #'s	Excellent	Mint Value	Rarity
#1 A	*Too Rare to Value	*Too Rare to Value	5
#1 A1	*Too Rare to Value	*Too Rare to Value	5
#1 B	*Too Rare to Value	*Too Rare to Value	5
#1 C	*Too Rare to Value	*Too Rare to Value	5
#1 D	$7,000	*Too Rare to Value	5
#1 E	$5,000	*Too Rare to Value	4
#1 F	*Too Rare to Value	*Too Rare to Value	5

*Box for 1932/33 Tall Model access #1 A & 2 A.
　　　　　　　　　　　　　　*Too Rare to Value　　5
(*The Zippo museum doesn't even own an example of this box. It is worth as much or more than the lighter that goes in it.)

*Lighters #2 A-#2 C must have totally original parts to be worth these prices with the exception of #2 A1, that is already noted in the text description.

Access #'s	Excellent	Mint Value	Rarity
#2 A	*Too Rare to Value	*Too Rare to Value	5
#2 A1	*Too Rare to Value	*Too Rare to Value	5
#2 B	*Too Rare to Value	*Too Rare to Value	5
#2 C	*Too Rare to Value	*Too Rare to Value	5
#2 D	$7,000	——	5
#2 E	$5,000	——	4
#3 A	$5,000	*Too Rare to Value	5
#3 B	$5,000	*Too Rare to Value	5
#3 C	$5,500	*Too Rare to Value	5
#4 A	$2,800	$5,000	5
#4 B	$3,000	$5,000	5
#5 A	$3,200	$5,500	5
#5 B	$3,000	$5,000	5
#5 C	$15,000	$18,000	5
#5 D	$6,000	$8,000	5
#5 E	$6,000	$8,000	5
#5 F	$6,000	$8,000	5
#5 G	$6,000	$8,000	5
#5 H	$4,500	$6,000	5

Access #'s	Excellent	Mint	Rarity Value
#5 I	$7,000	$10,000	5
#6 A	$3,500	$7,000	5
#6 B	$3,600	$7,000	5
#6 C	$3,700	$7000	5
#6 D	$12,000	$15,000	5
#6 E	$6,000	$8,000	5
#6 F	$6,000	$8,000	5
#6 G	$6,000	$8,000	5
#6 H	$4,500	$6,000	5
#6 I	$4,500	$6,000	5
#6 J	$12,000	$16,000	5

Chart 2: Rarity Values

*Important note: Beginning with Chart 2 and continuing through the remaining charts, Rarity Value only refers to mint never lit examples, unless otherwise noted in the text description.

5 = Less than 15 known to exist (Extremely Rare)
4 = Between 15 and 50 known to exist (Very Rare)
3 = Between 50 and 100 known to exist (Special Order/Rare)
2 = Regular Production Lighter/Unknown # (Uncommon)
1 = Regular Production Lighter (Common)

Quality Values

Mint: The lighter is new and has never been lit as well as having no mishandling flaws.
Excellent: The lighter has 100% original paint, if the lighter has an illustration, as well as having 100% chrome or nickel silver finish.

Access #'s	Excellent	Mint	Rarity Value
#7 A1	$7,500 (nearly mint)	$10,000	5

(The primary value of #'s 7 A1-#7 A4 is in the prototype case.)

#7 A2	$7,000 (used)	—	
#7 A3	$7,500 (nearly mint)	$10,000	5
#7 A4	$6,000 (used)	—	

VALUES for #8 B-#8 E don't include the value of having a piston insert, or the rarity and desirability of a specific advertising logo unless otherwise noted in the text description.

#7 B	$3,000	$5,000	5
#7 C	*Too Rare to Value	*Too Rare to Value	5
#7 D	*Too Rare to Value	*Too Rare to Value	5
#7 E	*Too Rare to Value	*Too Rare to Value	5
#7 F	*Too Rare to Value	*Too Rare to Value	5
#7 G	*Too Rare to Value	*Too Rare to Value	5
#7 H	*Too Rare to Value	*Too Rare to Value	5

VALUES for #8 B-#12 H don't include the value of a very rare and desirable advertising logo unless otherwise noted in the text description.

#8 A1	$1,000	$1,500	5
#8 A2	$900	—	
#8 A3	$800	$1,200	5
#8 A4	$400	—	

VALUES for #8 B-#12 H don't include the value of having a piston insert, or the rarity and desirability of a specific advertising logo unless otherwise noted in the text description.

Access #'s	Excellent	Mint	Rarity Value
#8 B	$1,000	$1,500	5
#8 C	$7,500	$10,000	5
#8 D	$1,000	$1,500	5
#8 E	$1,300	$1,800	5
#8 F	$5,000	$7,500	5
#9 A	$650	$1,000	5
#9 B	$850	$1,000	5
#9 C	$1,500	$2,000	5
#9 D	$5,000	*Too Rare to Value	5
#9 E	$5,000	*Too Rare to Value	5
#9 F	$850	$1,250	5
#9 G	$1,000	$1,500	5
#9 H	*Too Rare to Value	*Too Rare to Value	5
#9 I (K1, K2, K5, & K7) All four are gold plated brass prototypes.			
	$10,000	$15,000	5
#9 J	$850	$1,250	5
#9 K	*Too Rare to Value	*Too Rare to Value	5
#9 L	*Too Rare to Value	*Too Rare to Value	5
#9 M	$3,500	$4,500	5
#9 N	$850	$1,250	5
#9 O	Only one known to exist.	*Too Rare to Value	5
#9 P	Only one known to exist.	*Too Rare to Value	5
#10 A	$650	$1,000	5
#10 B	$850	$1,000	5
#10 C1	$850	$1,250	5
#10 D	$1,000	$1,500	5
#10 E	$1,500	$2,500	5
#10 F	$650	$850	5
#11 A	$650	$1,000	5
#11 B	$850	$1,000	5
#11 C1–C2	$850	$1,250	5
#11 C3	$1,000	$1,500	5
#11 C4–C9	$850	$1,250	5
#11 D	$1,000	$1,500	5
#11 E	$5,000	*Too Rare to Value	5
#11 F	*Too Rare to Value	*Too Rare to Value	5
#11 G (K1, K2, K5, & K7)			
	$5,000	$7,000	5
#11 H	$750	$1250	5
#11 I	*Too Rare to Value	*Too Rare to Value	5
#11 J	*Too Rare to Value	*Too Rare to Value	5
#11 K	$3,500	$5,000	5
#11 L	$1,000	$1,500	5
#11 M	$4,500	$6,000	5
#12 A	$650	$1,000	5
#12 B	$1,500	$2,000	5
#12 C	$1,500	$2,000	5
#12 D	$850	$1,000	5
#12 (E1–E9)	$850	$1,250	5
#12 F	$1,000	$1,500	5
#12 G	*Too Rare to Value	*Too Rare to Value	5
#12 H	$650	$850	5
#12 I	$850	$1,250	5
#12 J	$650	$850	5
#12 K	*Too Rare to Value	*Too Rare to Value	5

*(Only one known to exist with anywhere near this many illustrations this old.)

Numbers #13 A-#13 B1 must have the 1937-41 insert to be worth these values.

#13 A	$850	$1,250	5
#13 B–B1	$850	$1,250	5

World War II Black Crackles are valued and graded according to the amount of the black crackle paint still left on the surface of the lighter as well as the desirability of the illustration. As stated before, all lighters must have totally original inserts and hinges to even be considered at these prices.

Chart 3: Quality Values Numbers 14 A - #14 H2

Mint: The lighter has not been used in any way and has 100% original surface "crackle" paint. (If there are surface scratches or dents in the paint they should be taken into account when determining the value of the lighter.)

Excellent: The lighter has 100% original surface paint. The lighter has been used but has 100% paint.

Access #'s	Excellent	Mint	Rarity Value
#14 A	$450	$750	5
#14 B	$550	$750	5

#14 C is difficult to detect if in mint condition.

#14 C	$350	—	
#14 D	$500	$750	
#14 E	*Extremely rare and difficult to find in any condition. Valued @ $800		
#14 F	*Prices vary depending on what is on the lighter.		
#14 G	$500	$700	5
#14 H1	$700	$900	5
#14 H2	$1,200	$1,500	5

Chart 4: Quality Values for Access Numbers 15 A - #26 G1

Mint: The lighter is new and has never been lit as well as having no major mishandling flaws.

Excellent: The lighter has 100% original paint, if the lighter has an illustration, as well as having 100% chrome.

Access #'s	Excellent	Mint	Rarity Value
#15 A	$5,000	$7,500	5
#15 B	$5,000	$7,500	5

VALUES for #15 A & 15 B don't include the value of having a piston insert.

#16 A	$4,500	$6,500	5
#16 B	$4,500	$6,500	5
#17 A	$350	$500	5
#17 B	$350	$500	5
#17 C	$400	$600	5
#17 D	$350	$500	5
#17 E	$350	$500	5
#17 E1	$1,000	$1,500	5
#17 F	$350	$500	5
#17 G	$350	$500	5
#17 H	$350	$500	5
#17 I	$350	$500	5
#17 J	$350	$500	5
#17 K	$350	$500	5
#17 L	$350	$500	5
#17 M	$350	$500	5
#17 M1	$350	$500	5
#17 N	$350	$500	5
#17 O	$350	$500	5
#17 P	$350	$500	5
#17 Q	$350	$500	5
#17 R	$750	$1,000	5
#17 S	$400	$600	5
#17 T	$400	$600	5
#17 U	$350	$500	5
#17 V	$400	$600	5
#17 W	$350	$500	5
#17 X	$350	$500	5
#17 Y	$350	$500	5
#17 Z	$400	$600	5
#17 ZA	$350	$500	5
#17 ZB	$350	$500	5
#17 ZC	$1,000	$1,500	5
#17 ZD	$1,000	$1,500	5
#17 ZE	$3,500	$4,500 *If this motif is on a 1938-39 model.	5
#17 ZF	$350	$500	5
#17 ZG	$3,500	$5,000	5
#17 ZH	$4,500	$6,000 *If this motif is on a 1940-41 model.	5
#18 A	$250	$350	3
#18 B	$250	$350	3
#18 C	$250	$350	3
#18 C1	$250	$350	4
#18 D	$250	$350	3
#18 E	$250	$350	3
#18 F	$250	$350	5
#18 G	$500	$700	5
#18 H	$250	$350	4
#18 I	$250	$350	3
#18 I1 (without box)	$250	$350	5
(with box)	$350	$500	5
#18 J	$250	$350	3
#18 K	$250	$350	3
#18 L	$250	$350	5
#19 A	$250	$350	3
#19 B	$250	$350	3
#19 C	$250	$350	3
#19 D	$250	$350	3
#20 A	$500	$700	5
#20 B	$125	$200	2
#20 C	$125	$200	2
#20 D	$125	$200	2
#20 E	$125	$200	2
#20 F	$500	$700	5
#20 G	$700	$1,000	5
#20 H	$500	$700	5
#20 I	$500	$700	5
#20 I1	$350	$500	5
#20 J	$700	$1000	5

#20 K *There was only one test model of this motif ever produced. Valued @ $2,500.

#20 L	$125	$200	2
#20 M	$125	$200	2
#20 N	$1,500	$2,000	5
#20 O	$2,000	$2,500 (double-sided prototype)	5

#20 P	(#20 P is the reverse of #20 O. If by itself, would be $1,500-$2,000.)		5
#21 A	$2,500	$3,000	5
#21 B	$2,500	$3,000	5
#21 B1	$2,500	$3,000	5
#21 C	$2,500	$3,000	5
#21 C1	$2,500	$3,000	5
#21 D	$2,500	$3,000	5
#21 D1	$2,500	$3,000	5
#21 D2	$2,500	$3,000	5
#21 E	$2,500	$3,000	5
#21 F	$2,500	$3,000	5
#21 G	$2,500	$3,000	5
#21 H	$2,500	$3,000	5
#21 I	$2,500	$3,000	5
#22 A	$75	$125	2
#22 B	$75	$125	2
#22 C	$75	$125	2
#22 D	$75	$125	5
#22 E	$75	$125	2
#22 F	$350	$500	2
#22 G	$75	$125	2
#22 H	$75	$125	2
#22 I	$75	$125	2
#23 A	$150	$200	3
#23 B	$150	$200	3
#23 C	$150	$200	3
#23 D	$150	$200	3
#23 D1	$150	$200	3
#23 E	$150	$200	3
#23 F	$150	$200	3
#23 G	$150	$200	3
#24 A	$25	$35	2
#24 B	$25	$35	2
#24 C	$25	$35	2
#24 D	$25	$35	2
#24 E	$25	$35	2
#24 F	$25	$35	2
#24 F1	$700	$1,000	5
#24 G	$25	$35	2
#24 H	$700	$1,000	5
#24 I	$40	$60	2
#24 J	$700	$1,000	5
#24 K	$700	$1,000	5
#24 L	$700	$1,000	5
#24 M	$700	$1,000	5
#25 A	$25	$35	1
#25 A1	$700	$1,000	5
#25 B	$25	$35	1
#25 B1	$700	$1,000	5
#25 B2	$700	$1,000	5
#25 C	$25	$35	1
#25 C1	$300	$500	5
#25 D	$50	$75	3
#25 E	$25	$35	1
#25 E1	$700	$1,000	5
#25 F	$25	$35	1
#25 G	$25	$35	1
#25 H	$25	$35	1
#25 I	$25	$35	1
#25 J	$700	$1,000	5
#26 A	$25	$35	1
#26 B	$25	$35	1
#26 B1	$700	$1,000	5
#26 C	$500	$700	1
#26 D	$25	$35	1
#26 D1	$700	$1,000	5
#26 E	$25	$35	1
#26 F	$25	$35	1
#26 G	$25	$35	1
#26 G1	$700	$1,000	5

Town and Country illustrations will be graded and valued, for the most part, by the amount of "original paint" still found in the illustration.

Note that three barrel hinge 2032695 patents bring at least 50% more than five barrel hinge 2032695 patents which is true of all lighters, not just Town and Country models.

Chart 5: Quality Values Numbers 27 A - #27 A1

Mint: A lighter having not been used in any way and having 100% original surface color. (If there are surface scratches in the chrome they should be taken into account when determining the value of the lighter.)

Excellent: A lighter having 100% original surface color. The lighter is used but has 100% chrome.

Access #'s	Excellent	Mint	Rarity Value
#27 A	$1,200	$1,500	4
#27 B	$1,200	$1,500	4
#27 C	$1,200	$1,500	4
#27 D	$1,400	$1,700	4
#27 D1	$1,200	$1,500	5
#27 E	$1,200	$1,500	4
#27 F	$1,200	$1,500	4
#27 G	$1,200	$1,500	4
#27 G1	$1,800	$2,000	5
#27 H	$1,400	$1,700	5
#27 I	$2,500	$3,000	5
(the Marlin is the rarest of all Town & Country production lighters)			
#27 J	$2,500	$3,000	5
#27 K	$2,500	$3,000	5
#27 L	$2,500	$3,000	5
#27 M	$3,000	$3,500	5
#27 N	$3,000	$3,500	5
#27 O	$2,500	$3,000	5
#27 P	$2,500	$3,000	5
#27 Q	$3,000	$3,500	5
#27 R	$3,000	$3,500	5
#27 S	$2,500	$3,000	5
#27 T	$2,500	$3,000	5
#27 U (slim)	$1,200	$1,500	5
#27 U (reg.)	$1,500	$1,700	5
#27 V	$1,200	$1,500	5
#27 W	$1,200	$1,500	5
#27 X	$1,200	$1,500	5
#27 Y	$2,500	$3,000	5
#27 Z	$2,500	$3,000	5

Chart 6

World War II Black Crackles are valued and graded according to both the amount of original black crackle paint still left on the surface of the lighter as well as the desirability of the illustration.

Quality Values Numbers 28 A - #28 G:

Mint: A lighter having not been used in any way and having 100% original surface "crackle" paint. (If there are surface scratches or dents in the paint they should be taken into account when determining the value of the lighter.)
Excellent: A lighter having 100% original surface paint. The lighter has been used but has 100% paint.

Rarity value again refers only to mint, never lit, examples.

Access #'s	Excellent	Mint	Rarity Value
#28 A	$400	$600	3
#28 B	$400	$600	3
#28 C	$1,000	$1,500	4
#28 D	*Impossible to value unless sold at auction.		
#28 E	$400	$600	3
#28 F	*Hard to value due to the questionable authenticity of the emblems and the illustration(s). Original examples are extremely valuable and desirable!		
#28 G	$500	$700	4

Chart 7

"Regular production" models, from this point on, do not have a black crackle finish, although some prototypes do, There are Zippos, made in 1990's that have a type of black crackle finish.

Quality Values Numbers 28 H - #39 B:

Mint: The lighter is new and has never been lit as well as having no major mishandling flaws.
Excellent: The lighter has 100% original paint (if the lighter has an illustration as well as having 100% chrome).

Access #'s	Excellent	Mint	Rarity Value

This model lacks black crackle paint but still has the slightly curved out bottom.

#28 H	$300	$500	5
#29 A	$500	$700	3
#29 B	$500	$700	3
#30 A	$175	$250	2
#30 B	$175	$250	2
#31 A	$150	$200	2
#31 B	$150	$200	2
#31 C	$800	$1,000	5
#31 D	*Only one known to exist in any condition. Valued @ $5,000 in mint condition.		
#31 D1	$100	$150	4
#32 A	$100	$150	2
#32 B	$100	$150	2
#32 C	$100	$150	2

The advertising value of "rare and desirable" illustrations isn't taken into consideration on access #'s 33-42 E. The value, I ascertained, is for generic advertising logos; not those internationally known.

Starting with 33 B1 all models, to date, have some type of finish unless they are solid brass or are a prototype. The advertising motif, if any, which is on the surface of the lighter plays the largest role in determining the value of lighters manufactured from 1947 to date.

#33 A1	$350	$500	2
#33 A2	$500	$700	2
#33 B1	$300	$500	2
#33 B2	$350	$550	2
#33 C1	$275	$350	2
#33 C2	$300	$400	2

The value of having a lanyard is "about" 50% more than the value of those without the loss-proof lanyard, determined by production year in which they were made. There may be some exceptions. Therefore, if a regular model is $100, one with the lanyard would be in the neighborhood of $150, unless otherwise noted.

#33 D1	$500	$700	5
#33 D1A	$550	$750	5
#33 D2	$250	$350	4
#33 D2A	$200	$300	4
#33 D3	$150	$200	2
#33 D3A	$150	$200	2
#33 D3B	$150	$200	2
#34 A	$150	$200	2
#34 B	$150	$200	2
#34 C	$1,800	$2,500	5
#34 D-34 F	$175	$250	2
#34 G	$1,000	$1,500	2
#35 A	$250	$300	3
#35 B	$275	$325	3
#36 A	$150	$200	2
#36 B	$150	$200	2
#37 A	$150	$200	2
#37 B	$150	$200	2
#37 C	$1,000	$1,500	5
#38 A	$150	$200	2
#38 B	$150	$200	2
#39 A	$150	$200	2
#39 B	$150	$200	2

Chart 8: Quality Values Numbers 40 A1 - #41 B

Mint: Means the lighter is new and has never been lit as well as having no mishandling flaws.

Excellent: The lighter has 100% original leather and gold-leaf paint. If there is an illustration, 98% of the paint is there.

Access #'s	Excellent	Mint	Rarity Value
#40 A1-A4	$500	$700	3
#40 B	$700	$1,000	3
#41 A1-#41 A6	$350	$500	2
#41 B	$400	$600	3

Chart 9: Quality Values Numbers 42 A - #42 E and #43 G - #158 A

Mint: The lighter is new and has never been lit as well as having no major mishandling flaws.

Excellent: The lighter has 100% original paint (if the lighter has an illustration as well as having 100% finish).

Access #'s	Excellent	Mint	Rarity Value
#42 A	$150	$250	2
#42 B	$150	$250	2
#42 C	$150	$250	2
#42 D	$150	$250	3
#42 E	$500	$700	5

Important: The value of slim lighters is approximately 1/3 the value of regular size lighters in the same condition with the same motifs, although there are many exceptions. I will try to note as many exceptions as possible throughout the book. I will only discuss the 1956 slim and the 1958 14k gold slim at this time.

The primary value of the 1956 slim is the insert. Therefore I will give two values for the insert, those that lit and still have the wheel guard intact and those that are in mint condition, with all parts present.

#43 A	$300	$500	5
#43 B	$300	$500	5
#43 C	$700	$1,000	5
#43 D	$500	$700	5
#43D1	$200	$300	5
#43 E	$50	$75	5
#43 F	$300	$500	5
#43 G	$175	$250	4
#43 H	$175	$250	4
#43 I	$500	$500	5
#43 J	$200	$300	4
#44	$300	$500	3
#45 A	$400	$600	4
#45 B	$200	$300	2
#46 A	$300	$500	5
#46 B	$75	$100	3
#46 C	$75	$100	3
#46 D	$300	$500	5
#46 E	$300	$500	5
#46 F	$75	$100	3
#46 G	$75	$100	3
#46 H	$300	$500	5
#46 I	$75	$100	3
#47 A-#47 E (reg)	$150	$250	3
#47 A1-#47 E1 (slim)	$125	$175	3
#48 A-#48 C (reg)	$150	$250	3
#48 A1-#48 C1 (slim)	$125	$175	3
#49	$150	$250	4
#50	$200	$300	5
#51	$200	$300	5
#52 A	$250	$350	4
#52 B	$800	$1,000	5
#52 C	$250	$350	4
#52 D	$250	$350	4
#52 E	$150	$200	3
#52 F	$250	$350	4
#52 F1	$500	$700	5
#52 G	$250	$350	4
#52 H	$1,000	$1,500	5
#53 A	$400	$600	5
#53 B	$400	$600	5
#53 C	$700	$1,000	5
#54 A	$750	$1,000	5
#54 A1	$250	$350	5
#54 B	$250	$350	5
#54 C	$250	$350	5
#54 D	$100	$125	3
#54 E	$200	$250	4
#54 E1	$100	$150	4
#54 F	$75	$125	4
#54 F1	$50	$75	4
#54 G	$100	$125	4
#54 H	$75	$100	3
#54 H1	$60	$75	3
#55 A	$300	$400	1
#55 B	$300	$400	5
#55 C	$1,000	$1,500	5
#55 D	$400	$600	5
#55 E	$3,000	$3,500	5
#55 F	$150	$225	4
#55 G	$150	$225	4
#55 H	$100	$150	3
#55 I	$100	$150	5
#56 A-#56 L	$100	$150	2
#56 M	$3,000	$5,000	5
#56 N	$3,000	$5,000	5
#56 O	$3,000	$5,000	5
#56 P	$3,000	$5,000	5
#56 Q	$500	$700	5
#56 R	$500	$700	5
#56 S	$500	$700	5

Item	Low	High	Rarity
#56 T	$1,000	$1,500	5
#57 A	$100	$150	3
#57 B	$75	$125	2
#57 C	$50	$75	2
#57 C1	$50	$75	2
#57 D	$100	$150	3
#57 E	$100	$150	3
#57 F	$100	$150	3
#57 G	$50	$75	3
#58 A	$50	$75	2
#58 B	$150	$200	4

Note that #59 is a Key Holder and not a lighter.

Item	Low	High	Rarity
#59 A	$150	$200	5
#59 B	$15	$20	2
#60 A	$50	$75	2
#61 A	$150	$200	4
#61 B	$75	$100	2
#61 C	$75	$100	2
#61 D	$75	$100	2
#61 E	$40	$60	2
#61 F	$40	$60	2
#61 G	$40	$60	2
#62 A	$75	$125	2
#62 B	$40	$60	2
#62 C	$500	$700	5
#63 A	$50	$75	2
#64 A	$50	$75	2
#65 A	$1,000	$1,500	5
#65 B	$1,000	$1,500	5
#65 C	$1,000	$1,500	5
#66 A	$40	$60	2
#66 B-#66 B1	$50	$75	1
#66 C	$50	$75	2
#66 D	$350	$500	5
#66 E	$500	$700	5
#67 A	$50	$75	2
#67 B	$50	$75	2
#68 A	$50	$75	2
#68 A1	$500	$750	5
#68 A2	$500	$750	5
#68 B	$50	$75	2
#68 B1	$500	$750	5
#68 B2	$500	$750	5
#69 A	$100	$150	2
#70 A	$35	$50	2
#70 B	$1,000	$1,500	5
#71 A	$40	$60	2
#71 B	$20	$40	2
#71 C	$100	$125	2
#72 A	$25	$30	2
#72 B	$35	$45	2
#73 A-73 E	$20	$25	1
#73 F	$25	$35	2
#74 A	$50	$75	2
#74 B	$100	$150	5
#75 A	$35	$50	2
#75 B	$35	$50	2
#75 C	$35	$50	2
#75 D-#75 G	$25	$30	2
#75 H	$750	$1,000	5
#75 H1	$750	$1,000	5
#75 I	$500	$700	5
#75 J	$1,500	$2,000	5
#76 A-#76 H	$20	$25	1
#77 A-#77 F	$20	$25	1
#77 B1	$300	$500	5
#78 A-#78 F	$20	$25	2
#78 B1	$400	$500	5
#79 A-#79 D	$25	$30	2
#80 A	$25	$30	2
#80 A1	$300	$500	5
#80 A2	$300	$500	5
#80 B	$25	$30	2
#80 C	$25	$30	2
#80 C1	$300	$500	5
#80 D	$25	$30	2
#80 D1	$200	$400	5
#81 A-#81 H	$25	$30	1
#82 A-#82 D	$25	$30	1
#83 A	$60	$80	1
#83 B Mint in the original box; Valued @ $300-$400.			
#83 C	$1,500	$2,000	5
#83 D	$1,500	$2,000	5
#83 E	$1,500	$2,000	5
#84 A-#84 F	$25	$30	1
#84 G	$750	$1,250	5
#84 H	$200	$300	2
#84 I	$100	$150	1
#85 A-#85 D	$25	$30	1
#85 E-#85 H	$500	$750	5
#86 A-#86 H	$25	$30	1
#87 A-#87 H	$25	$30	1
#88 A	$60	$100	1
#89 A-#89 H	$25	$30	1
#89 B1 & #89 D1	$500	$750	5
#90 A-#90 H	$25	$30	1
#91 A-#91 D	$25	$30	1
#92 A-#92 F	$25	$30	1

Access #'s	Excellent	Mint	Rarity Value
#93 A-#93 F	$25	$30	1
#94 A-#94 F	$25	$30	1
#94 G-#94 N	$25	$30	1
#95 A-#95 D	$25	$30	1
#96 A-#96 G	$25	$30	1
#97 A-#97 D	Zippo retails them for $63.95.		1
#97 E-#97 F	$400	$600	5
#98 A-#98 H	$25	$30	1
#99 A-#99 G	$25	$30	1
#99 G1	$200	$300	5
#100 A-#100 C	$35	$50	1
#101 A-#101 J	$25	$30	1
#102 A-#102 G	Zippo retails them for $52.95.		1
#103 A-#103 F	$25	$30	1
#104 A	$25	$30	1
#105 A	$25	$30	1
#106 A-#106 D	$30	$40	1
#106 C1	$400	$600	5
#107 A-#107 C	$30	$40	1
#108 A	$30	$40	1
#109 A-#109 D	$25	$30	1
#110 A-#110 D	$30	$40	1
#111 A-#111 E	$25	$30	1
#112 A	Zippo retails these for $3,110.95.		1
#112 B	Zippo retails these for $2,158.95.		1
#113 A-#113 F	$25	$30	1
#113 G-#113 N	$30	$40	1
#114 A-#114 F	Zippo retails them for $52.95.		1
#115 A-#115 C	$25	$30	1
#116 A-#116 F	$25	$30	1
#117 A-#117 G	$25	$30	1
#118 A-#118 E	$25	$30	1
#119 A-#119 G	$25	$30	1
#120 A-#120 D	$25	$30	1
#121 A-#121 C	$25	$30	1
#122 A-#122 C	$25	$30	1
#123 A-#123 C	$25	$30	1
#124 A1-#124 A4	$25	$30	1
#125 A-#125 H	$25	$30	1
#126 A-#126 N	$25	$30	1
#127 A-#127 C	$25	$30	1

#128- #156 1996 lighters can be purchased from Zippo at prices in the text.

Access #'s	Excellent	Mint	Rarity Value
#157 A	$5,000	$8,000	5
#157 B	$700	$1,000	5
#157 C	$2,500	$3,500	5
#158 A	$150	$250	2

Chart 10: Quality Values Numbers 158 B1 - #158 B9 and #159 B1 - #159 B9

Use Chart 9 for #158 C-D1, 159 A, & #159 C-E

Mint: The lighter has not been used in any way having 100% original surface color. (If there are surfaces scratches in the chrome they should be taken into account when determining the value of the lighter.)

Excellent: The lighter has 100% surface color. The lighter is used but has 100% chrome.

Access #'s	Excellent	Mint	Rarity Value
#158 B1	$3,000	$3,500	5
#158 B2	$4,000	$5,000	5
#158 B3	$3,000	$3,500	5
#158 B4	$4,000	$5,000	5
#158 B5	$3,000	$3,500	5
#158 B6	$3,000	$3,500	5
#158 B7	$3,000	$3,500	5
#158 B8	$3,000	$3,500	5
#158 C	$150	$250	5
#158 D	$500	$700	5
#158 D1	$600	$800	5
#158 E	$500	$700	5
#159 A	$150	$250	2
#159 B1	$1,500	$2,500	5
#159 B2	$4,000	$5,000	5
#159 B3	$1,500	$2,500	5
#159 B4	$4,000	$5,000	5
#159 B5	$1,500	$2,500	5
#159 B6	$1,500	$2,500	5
#159 B7	$1,500	$2,500	5
#159 B8	$1,500	$2,500	5
#159 B9	$4,000	$5,000	5
#159 C	$250	$300	2
#159 D-#159 D1	$450	$600	5
#159 E	$250	$400	5

Use Chart 8 for access #159 F-#159 I.

#159 F	$3,500	$5,000	5
#159 G	Only one is known to exist.		
	Valued @ $5,000-$7,000.		5
#159 H	Only three are known to exist.		
	Valued @ $5,000-$7,000.		5
#159 I	$250	$400	5

Use Chart 9 for access #160 A-#162 A.

#160 A	$7,000	$9,000	5
	(Extremely rare!)		
#161 A	$250	$350	2
#162 A	$75	$125	2

Use Chart 10 for access #162 B-E.

#162 B	Only one known to exist.		
	Valued @ $20,000-$25,000+.		5

*This is the finest Town & Country illustration I've ever seen and quite possibly the best illustration that Zippo has produced since 1932! This lighter is almost too rare and desirable to even be valued!

#162 C	$1,500	$2,500	5
#162 D	$1,000	$1,500	5
#162 E	$1,000	$1,500	5

Use Chart 9 for access #162 F-#165 D.

#162 F	$450	$600	5
#162 G	$750	$1,000	5
#162 H	$150	$250	2
#162 I	Test model, only one know to exist.		
	Valued @ $1,500-$2,000.		5
#162 J	$125	$150	3
#162 K	$175	$250	3
#162 L	$175	$250	3
#162 M	$175	$250	3
#162 N	$125	$150	3
#163 A	$200	$250	3
#163 B	$250	$350	3
#164 A	$350	$500	3
#164 B	$250	$350	3
#164 C	$175	$250	3
#164 D	Prototype, only one known to exist.		
	Valued @ $3,500-5,000.		5
#165 A	$75	$100	1
#165 B	$100	$150	2
#165 C	$350	$500	5
#165 D	$350	$500	5

Chart 11: Quality Values Numbers 166 - #174 C:

Chart #11 deals with Zippo accessory Items therefore I will only give two quality values. The first will represent an item in excellent condition. The second value will represent an item that is brand new.

Access #'s	Excellent	Mint
#166	$15	$20
#167 A	$500	$700
#167 B	$50	$75
#167 C	$15	$20
#167 D	$35	$50
#167 E	$20	$30
#167 F	$150	$200
#168 A	$15	$20
#168 B	$35	$50
#169 A	$35	$50
#169 B	$40	$60
#170 A	$15	$20

I didn't cover in detail nor value #171 in the text.

#172 A	$15	$20
#172 B	$15	$20
#173 A	$500	$700
#173 B	$50	$75

Chart 12: Flint Dispensers

Flint dispenser illustrations are found on the chart on page 141 in the text. Access numbers are taken from the chart on that page. I only listed one quality value for flint dispensers. Excellent condition dispensers are complete although the flint may have oxidized and be powder inside.

	Excellent Condition
#174 A	$125
#174 B	$75
#174 C	$60
#174 D	$60
#174 E	$60
#174 F	$125
#174 G	$75
#174 H	$75
#174 H1	$100
#174 H2	$125
#174 I	$30
#174 J	$20
#174 K	$30
#174 L	$35
#174 L1	$100
#174 M	$3

#174 N	$1
#174 O	$10
#174 P	$10
$174 Q	$3
#174 R	$3
#174 S	$1
#174 T	$1
#174 U	$350

Chart 13: Lighter Fluid Cans

I only listed one quality value for lighter fluid cans. Containers in excellent condition must have 99% paint and/or paper.

	Excellent Condition
#175 A	$700
#175 B	$600
#175 C	$600
#175 D	$600
#175 E	$600
#175 F	$150
#175 G	$125
#175 H-#175 J	$100
#175 K	$300
#175 L-#175 O	$100
#175 P-#175 Q	$125
#175 R-#175 V	$50
#175 W-#175 Z	$100
#175 A1	$500
(few survived)	
#175 A2-#175 A3	$20
#175 A4	$40
#175 A5-#175 A14	$20
#175 A15	$40
#175 A16-#175 A24	$10
#175 A25	$40
#175 A26-#175 A33	$5
#175 A34	$3
#175 A35-#175 A43	$2

Chart 14: Wick Displays

I only listed one quality value for wick displays. Displays in excellent condition cannot be ripped or torn, although some of the packets may be missing.

	Excellent Condition
#176 A	$200
#176 B	$60
#176 C	$200

Chart 15: Quality Values for Numbers 177 A - #188 D

Chart 15 deals with specific illustrations that were done in a myriad of different processes. Due to the difficulty in grading them as I did all previous lighters I just listed them in two categories; excellent condition and mint condition. I listed no rarity value.

Access #'s	Excellent	Mint
#177 A	$800	$1,000
#177 B	$600	$800
#177 C	$600	$800
#177 D	$600	$800
#177 E	$75	$125
#177 F	$175	$250
#177 F1	$175	$250
#177 G	$175	$250
#177 H	$175	$250
#177 I	$175	$250
#177 I1	$175	$250
#177 J	$500	$700
#177 J1-#177 J3	$500	$700
#177 J4	$300	$400
#177 J5	$300	$400
#177 J6	$500	$700
#177 J7	$800	$1,200
#177 K	$150	$200
#177 L	$150	$200
#177 M	$150	$200
#177 N	$100	$150
#177 N1	$200	$300
#177 O	$100	$150
#177 O1	$200	$300
#177 P	$200	$300
#177 Q	$100	$150
#177 Q1	$200	$300
#177 R-#177 S	$100	$150
#177 T-#177 V	$35	$50
#177 W-#177 Y	$20	$35
#178 A	$1,000	$1,200
#178 A1	$1,200	$1,500
#178 B	$1,200	$1,500
#178 C	$400	$600
#178 C1	$300	$500
#178 D1	$1500	$2000
#178 D2	$1200	$1500
#178 D3	$250	$350
#178 E	$18,000	$22,000
#178 F	$400	$600
#178 F1	$400	$600
#178 G1	$500	$700
#178 G2	$1,000	$1,500
#178 G3	$125	$150
#178 G4	$50	$75
#178 G5	$75	$100
#178 G6	$85	$100
#178 G7	$35	$50
#178 G8	$35	$50
#178 G9	$3,000	$3,500
#178 G10	$50	$75
#178 G11	$400	$600
#179 A	$400	$500

#		
#179 B	$1,000	$1,200
#179 C	$20	$30
#179 D	$20	$30
#180 A	$300	$400
#180 B	$800	$1,200
#180 C	$300	$400
#180 D	$300	$400
#180 E	$500	$700
#180 F	$300	$400
#180 G	$800	$1,200
#180 H	$800	$1,200
#180 I	$800	$1,200
#180 J	$1,500	$2,000
#180 K	$1,500	$2,000
#180 L	$1,200	$1,500
#181 A	$75	$125
#181 B	$250	$350
#181 C	$50	$75
#181 D	$350	$500
#182 A	$350	$500
#182 B	$350	$500
#183 A	$250	$350
#183 B	$500	$700
#184 A	$250	$300
#184 B	$250	$350
#184 C	$250	$350
#184 D	$200	$300
#184 E	$35	$50
#184 F	$250	$300
#184 G	$100	$150
#184 H	$500	$700
#184 I	$150	$200
#185 A	$800	$1,200
#185 B	$800	$1,200
#185 C	$800	$1,200
#185 D	$800	$1,200
#185 E	$800	$1,200
#185 F	$800	$1,200
#186 A	$800	$1,200
#187 A	$200	$300
#187 B	$150	$200
#187 C	$350	$500
#187 D	$350	$500
#187 E	$100	$150
#187 F	$60	$75
#187 G	$400	$600
#187 H	$800	$1,200
#187 I	$500	$700
#187 J	$500	$700
#187 K	$600	$800
#187 L	$600	$800
#187 M	$150	$250
#187 N	$100	$150
#187 O	$100	$150
#187 P	$800	$1,200
#187 Q	$200	$300
#187 R	$800	$1,200
#187 S	$350	$500
#187 T	$400	$600
#187 U	$150	$200
#187 V	$200	$300
#187 W	$350	$500
#187 X	$200	$300
#187 X1	$400	$600
#187 Y	$350	$450
#187 Z	$125	$175
#187 AA	$200	$300
#187 AB	$250	$350
#187 AC1	$250	$350
#187 AC2	$250	$350
	(prototype)	
#187 AC3	$75	$100
	(production model)	
#187 AD	$300	$400
#187 AE	$350	$500
#187 AF	$800	$1,200
#187 AG	$350	$500
#187 AH	$750	$1,000
#187 AI	$50	$75
#187 AJ	$250	$350
#187 AK	$350	$500
#187 AL	$35	$50
#187 AM	$250	$300
#187 AN	$175	$225
#187 AO	$300 ea.	$400 ea.
#187 AP	$75	$125
#187 AQ	$500	$700
#187 AR	$75	$125
#187 AS	$35	$50
#187 AT		
-#187 AU	$400	$600
#187 AV	$150	$200
#187 AW	$150	$200
#188 A	$400	$600
#188 B	$750	$1,000
#188 C	$500	$700
#188 D	$500	$700

Chart 16: Quality Values for Numbers 189 A – #189 G

I only listed one quality value for the engraving plates. These were used in conjunction with a panograph.

Access #'s	Excellent Condition
#189 A-#189 H	$150 ea.

Chart 17: Quality Values for Numbers 190 A - #191 A

Mint: The lighter is new and has never been lit as well as having no major mishandling flaws.

Excellent: The lighter has 100% original paint (if the lighter has an illustration as well as having 100% finish).

Access #'s	Excellent	Mint	Rarity Value
#190 A	$200	$275	3
#190 B	$200	$275	3
#190 C	$300	$500	3
#190 D	$300	$500	3
#191 A	$1,200	$1,500	5

I only listed one value for the dash board lighter holder. The item must be in excellent condition.

Access #'s	Excellent Condition
#192 A	$200

I only listed one value for the Tiffany Bamboo lighter. The item must be in excellent condition.

Access #'s	Excellent Condition
#193 A	$2,000

I only listed one value for access #'s 194 A-#196 A. The item must be in excellent condition.

Access #'s	Excellent Condition
#194 A	$100
#195 A	$85
#196 A	$300
#197 A	$200
#199 A	$5,000-$7,500

Bibliography

Friedel, Robert. *Zipper: An Exploration in Novelty*. New York: W. W. Norton & Co., 1994.

Schneider, Stuart and George Fischler. *Cigarette Lighters*. Atglen, PA: Schiffer Publishing Ltd., 1996.

NOTES